The Skeletal and Muscular Systems

YOUR BODY How It Works

The Skeletal and Muscular Systems

Gregory J. Stewart
Bureau of Arms Control
U.S. Department of State
Washington, D.C.

Introduction by
Denton A. Cooley, M.D.
President and Surgeon-in-Chief
of the Texas Heart Institute
Clinical Professor of Surgery at the
University of Texas Medical School, Houston, Texas

CHELSEA HOUSE
P U B L I S H E R S
A Haights Cross Communications Company
Philadelphia

CHELSEA HOUSE PUBLISHERS
VP, NEW PRODUCT DEVELOPMENT Sally Cheney
DIRECTOR OF PRODUCTION Kim Shinners
CREATIVE MANAGER Takeshi Takahashi
MANUFACTURING MANAGER Diann Grasse

Staff for THE SKELETAL AND MUSCULAR SYSTEMS
EXECUTIVE EDITOR Tara Koellhoffer
ASSOCIATE EDITOR Beth Reger
PRODUCTION EDITOR Megan Emery
PHOTO EDITOR Sarah Bloom
SERIES & COVER DESIGNER Terry Mallon
LAYOUT 21st Century Publishing and Communications, Inc.

A Haights Cross Communications ⬥ Company

www.chelseahouse.com

First Printing

1 3 5 7 9 8 6 4 2

Library of Congress Cataloging-in-Publication Data

Stewart, Gregory, 1957–
    The skeletal and muscular systems/Gregory J. Stewart.
        p. cm.—(Your body, how it works)
    ISBN 0-7910-7905-8
    1. Musculoskeletal system. I. Title. II. Series.
QP301.S83 2005
612.7—dc22

                                                        2004006569

Table of Contents

Introduction

The human body is an incredibly complex and amazing structure. At best, it is a source of strength, beauty, and wonder. We can compare the healthy body to a well-designed machine whose parts work smoothly together. We can also compare it to a symphony orchestra in which each instrument has a different part to play. When all of the musicians play together, they produce beautiful music.

From a purely physical standpoint, our bodies are made mainly of water. We are also made of many minerals, including calcium, phosphorous, potassium, sulfur, sodium, chlorine, magnesium, and iron. In order of size, the elements of the body are organized into cells, tissues, and organs. Related organs are combined into systems, including the musculoskeletal, cardiovascular, nervous, respiratory, gastrointestinal, endocrine, and reproductive systems.

Our cells and tissues are constantly wearing out and being replaced without our even knowing it. In fact, much of the time, we take the body for granted. When it is working properly, we tend to ignore it. Although the heart beats about 100,000 times per day and we breathe more than 10 million times per year, we do not normally think about these things. When something goes wrong, however, our bodies tell us through pain and other symptoms. In fact, pain is a very effective alarm system that lets us know the body needs attention. If the pain does not go away, we may need to see a doctor. Even without medical help, the body has an amazing ability to heal itself. If we cut ourselves, the blood clotting system works to seal the cut right away, and the immune defense system sends out special blood cells that are programmed to heal the area.

During the past 50 years, doctors have gained the ability to repair or replace almost every part of the body. In my own field of cardiovascular surgery, we are able to open the heart and repair its valves, arteries, chambers, and connections. In many cases, these repairs can be done through a tiny "keyhole" incision that speeds up patient recovery and leaves hardly any scar. If the entire heart is diseased, we can replace it altogether, either with a donor heart or with a mechanical device. In the future, the use of mechanical hearts will probably be common in patients who would otherwise die of heart disease.

Until the mid-twentieth century, infections and contagious diseases related to viruses and bacteria were the most common causes of death. Even a simple scratch could become infected and lead to death from "blood poisoning." After penicillin and other antibiotics became available in the 1930s and '40s, doctors were able to treat blood poisoning, tuberculosis, pneumonia, and many other bacterial diseases. Also, the introduction of modern vaccines allowed us to prevent childhood illnesses, smallpox, polio, flu, and other contagions that used to kill or cripple thousands.

Today, plagues such as the "Spanish flu" epidemic of 1918–19, which killed 20 to 40 million people worldwide, are unknown except in history books. Now that these diseases can be avoided, people are living long enough to have long-term (chronic) conditions such as cancer, heart failure, diabetes, and arthritis. Because chronic diseases tend to involve many organ systems or even the whole body, they cannot always be cured with surgery. These days, researchers are doing a lot of work at the cellular level, trying to find the underlying causes of chronic illnesses. Scientists recently finished mapping the human genome, which is a set of coded "instructions" programmed into our cells. Each cell contains 3 billion "letters" of this code. By showing how the body is made, the human genome will help researchers prevent and treat disease at its source, within the cells themselves.

The body's long-term health depends on many factors, called risk factors. Some risk factors, including our age, sex, and family history of certain diseases, are beyond our control. Other important risk factors include our lifestyle, behavior, and environment. Our modern lifestyle offers many advantages but is not always good for our bodies. In western Europe and the United States, we tend to be stressed, overweight, and out of shape. Many of us have unhealthy habits such as smoking cigarettes, abusing alcohol, or using drugs. Our air, water, and food often contain hazardous chemicals and industrial waste products. Fortunately, we can do something about most of these risk factors. At any age, the most important things we can do for our bodies are to eat right, exercise regularly, get enough sleep, and refuse to smoke, overuse alcohol, or use addictive drugs. We can also help clean up our environment. These simple steps will lower our chances of getting cancer, heart disease, or other serious disorders.

These days, thanks to the Internet and other forms of media coverage, people are more aware of health-related matters. The average person knows more about the human body than ever before. Patients want to understand their medical conditions and treatment options. They want to play a more active role, along with their doctors, in making medical decisions and in taking care of their own health.

I encourage you to learn as much as you can about your body and to treat your body well. These things may not seem too important to you now, while you are young, but the habits and behaviors that you practice today will affect your physical well-being for the rest of your life. The present book series, YOUR BODY: HOW IT WORKS, is an excellent introduction to human biology and anatomy. I hope that it will awaken within you a lifelong interest in these subjects.

Denton A. Cooley, M.D.
President and Surgeon-in-Chief
of the Texas Heart Institute
Clinical Professor of Surgery at the
University of Texas Medical School, Houston, Texas

1

The Skeletal and Muscular Systems:

The Movers and Shakers of the Human Body

INTRODUCTION

Recently, one of the major television networks featured a "strongman" competition. In one phase of the contest, the competitors balanced a metal beam across their knees and supported thousands of pounds of weight on their lower legs. On another occasion, several years earlier, I attended a performance of the American Ballet Theater to watch in awe, as the lead dancer, in performing a series of leaps and turns, appeared to defy gravity. In both cases my immediate response was, "That is impossible; how can they do that?" In this book we will learn the answers to these and many other mysteries. The skeletomuscular system, a combination of the body's skeletal system and muscular system, is responsible for these acts of strength, power, and grace. We will see that what at first appears impossible is probably better described as difficult but possible (Figure 1.1).

The secret to the heavy weight-balancing act is actually the incredible strength of the bones of the legs. By balancing the weight directly over the long line of the bones of the lower legs, and by not attempting to move the weight once in place, the

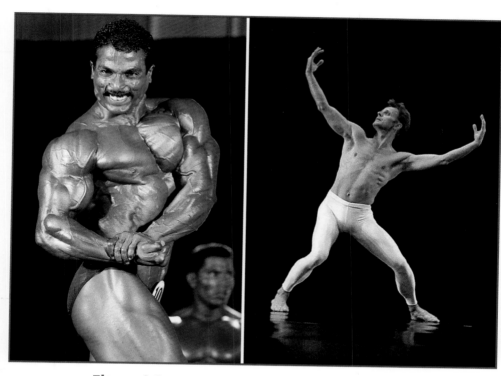

Figure 1.1 The skeletomuscular system is responsible for both the strength and grace of our movements. The man pictured on the left is a bodybuilder, and illustrates the strength our bodies can achieve. The man pictured on the right is a ballet dancer, who demonstrates the grace and precision of movement we can achieve.

"strongman" or "strongwoman" can support weights many times greater than his or her weight. As we will see, the ability to support extraordinary weight is an important feature of many of the bones of the body. The secret to the dancer who appeared to defy gravity lies in the incredible strength of the muscles of the legs. The dancer had strengthened and trained the muscles of his legs to propel him high off the stage and to complete the graceful turns and twists before his feet returned to the surface of the stage. These are dramatic examples of the power and precision of the skeletomuscular system, but the movements that our bones and muscles allow each of us to perform everyday are no less

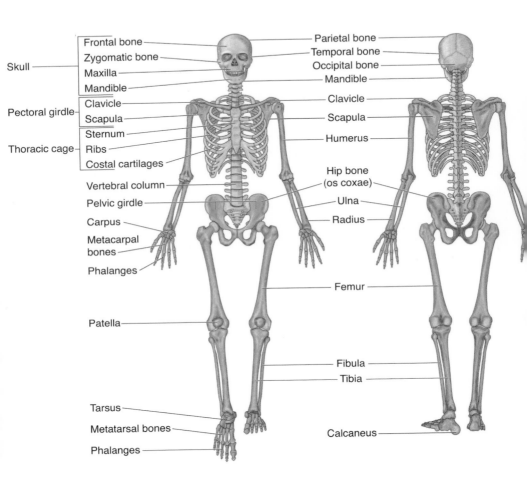

Figure 1.2 The human skeleton consists of 206 bones, which are organized along lines of symmetry for the body. The figure on the left is a frontal view and the figure on the right a rear view of the human skeleton.

amazing. As we move through this book, we will discover the importance of our skeleton and muscles and how these combined systems contribute to our ability to function in the world.

INSIDE VERSUS OUTSIDE ARMOR

The human skeleton (Figure 1.2), which is made up of 206 bones, is critical for positioning and stabilizing the other organ systems. Because our skeleton is inside the tissues of our body, it is referred to as an **endoskeleton**. Not all organisms have skeletons. Bacteria, protozoa, and fungi are all examples of living organisms that lack skeletons. These are all microscopic single-celled organisms. As organisms become more complex and increase in size, they develop the need for a skeleton of some type. Many insects and crustaceans (shellfish) have an **exoskeleton**, a rigid tough protective layer on their outside. Exoskeletons provide for strength and a degree of movement, but they have their limitations. For one thing, the growth of the organism is restricted in phases. After a certain amount of growth, the exoskeleton becomes constrictive and the organism must **molt**, or "shed" its exoskeleton, in order to become larger (Figure 1.3). During a molting phase, these organisms are particularly vulnerable to damage and to predators. Ultimately, organisms with exoskeletons are limited in the size they can achieve; consequently, there are no exoskeleton organisms among the largest animals on Earth or in its waters.

CRITICAL FUNCTIONS OF THE SKELETAL SYSTEM

The endoskeleton sacrifices the protection of a very tough coat for greater ability to grow and greater mobility. Most large animals have endoskeletons. Although the tough outer armor is missing in these organisms, the endoskeleton still provides protection of the organism and its internal organs. It does this by providing a solid framework

Figure 1.3 Insects are encased in a hard shell, or exoskeleton. As the insect grows, this exoskeleton becomes confining. In this photograph, an insect sheds its old exoskeleton, which has become too small. A new, larger exoskeleton will form after the insect completely sheds the original shell.

on which the rest of the body's tissues are attached. Consider what a mason does when making a large concrete structure. He or she incorporates steep reinforcement bars (rebar) into the frame for the concrete. As the concrete sets and becomes hard, the embedded rebar makes it much stronger. Our skeleton works in much the same way. In some cases, this general protection by the endoskeleton may not be enough. When organs of the body are especially critical and sensitive to damage (for example, the heart and brain), the skeleton has developed to surround these organs, providing specialized physical protection. Thus, the first critical function of a skeleton is to provide general and specialized protection.

THE FATE OF A SOFT-SHELL CRAB

While humans and other mammals have an endoskeleton, many organisms have exoskeletons—that is, a skeleton on the outside of the body, not on the inside. Most insects have exoskeletons. Beetles, for example, have a hard exoskeleton rich in a chemical called **chitin**. This makes the outer coat of the beetle very tough. In fact, if you have ever stepped on a beetle or a cockroach, you may have noticed that the insect often survives with no apparent damage. Crabs are oceanic animals that have exoskeletons. When a crab grows, it gets bigger and bigger until its rigid exoskeleton will not let it grow any further. At this point, the crab will shed its outer skeleton and make a larger one, allowing the crab to continue to grow in size. In the Chesapeake Bay area of Maryland, certain species of crabs shed their exoskeletons around May of each year. These are the famous Maryland soft-shell crabs. Immediately after they shed their tough coat, the crabs are extremely vulnerable to predators, particularly to people who make a living fishing for them. The soft-shell crabs are gathered and easily prepared. Because the tough outer layer is missing, the entire crab is edible. No mallets or cracking tools are needed to enjoy a delicious feast of soft-shell crabs. You just pop them in your mouth and chew.

A second important function of the skeleton is to provide resistance for the muscles. In order for muscles to help us move, they must have solid points to act against. The **skeletal muscles** are muscles that allow us to move from one place to another or allow us to change the position of parts of our body while we remain in the same place (Figure 1.4). The skeletal muscles attach their ends to the bones of the skeletal system. By contracting between two bones, the muscles change the position of those bones relative to each other, causing the body to move. A simple way to envision this is to take two pieces of wood that are joined with a hinge. If you put hooks in the ends of the pieces of wood that are away from the hinged ends and attach a rubber band to them, the two pieces of wood will snap together. If you stretch the rubber band without attaching it to the hooks, it simply goes back to its original shape without moving anything but itself. For muscles to do work, they must be attached to bones.

A third critical function of our skeleton is to facilitate other body functions. The jaw and the teeth are part of the skeletal system, but are essential to our digestive system because they begin the process of food digestion. Tiny bones in the ear are essential for transmitting vibrations that become sounds and are recognized by our brain. Finally, the **ribs** surround the lungs and create the chest cavity, which is enclosed by bone, muscle, and a special muscle called the diaphragm. By expanding the area of this closed chamber, air is pulled into the lungs. When the space is contracted, it forces the air out. Thus, the rigid nature of the ribs surrounding the chest cavity makes breathing possible.

A fourth essential function of the skeletal system is the production of other important cells. Inside certain bones is a soft tissue called **bone marrow**. Bone marrow cells produce the blood cells that are essential for transport of oxygen and carbon dioxide throughout the body and that are needed for the immune system.

CRITICAL FUNCTIONS OF THE MUSCULAR SYSTEM

The first critical function of the muscular system has been alluded to in the previous section. Skeletal muscles work in **opposing pairs** to move the skeleton and therefore move the body in part or in whole. As we will see later in this book, skeletal muscles only work by pulling on bones. They cannot push on them. As a result, skeletal muscles must work in pairs to provide full range of movement for a particular bone or **joint**, the point where two bones come together. Thus, the first critical function of the muscles is to direct **voluntary movement**, movement of a muscle or limb that is under conscious control.

A second critical function of muscles is communication. The muscles of the face, especially those of the jaw and tongue, are incredibly complex and finely controlled. The intricate nature of these muscles, along with those of the **larynx**, or **voice box**, allow us to make the multitude of sounds that lead to verbal communication. Persons who are mute (unable to speak) can also communicate through sign language, using the muscles of the hands and arms to create equally intricate and subtle movements. Without our muscular system, human communication as we know it would not be possible.

A third critical role for muscles is maintaining the body's vital functions without our awareness. Another group of muscles, the **involuntary muscles**, control essential bodily functions without the need for our conscious direction. The functions of circulation, respiration, and digestion are all controlled by involuntary muscles.

The fourth critical function for our muscular system is stabilization of the body. Although the skeleton provides rigid support for our body, the muscles, by balancing the pull of opposing pairs, act to stabilize the bones and therefore hold the body in place. Our posture is dependent on the stabilizing effect of our muscles. Attachment of muscles

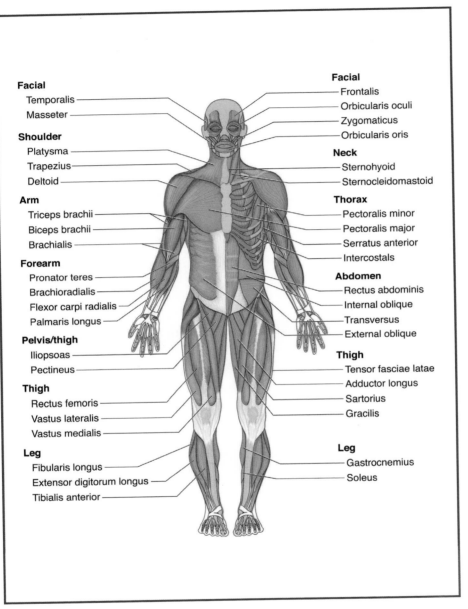

Figure 1.4 Frontal and rear views of the human body illustrate the muscular system. The muscles, along with the bones, provide us with support and allow us to move. The body has several

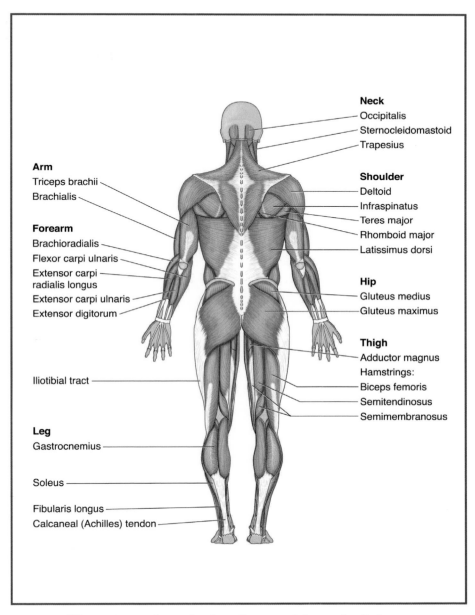

Neck
Occipitalis
Sternocleidomastoid
Trapesius

Arm
Triceps brachii
Brachialis

Forearm
Brachioradialis
Flexor carpi ulnaris
Extensor carpi
radialis longus
Extensor carpi ulnaris
Extensor digitorum

Shoulder
Deltoid
Infraspinatus
Teres major
Rhomboid major
Latissimus dorsi

Hip
Gluteus medius
Gluteus maximus

Thigh
Adductor magnus
Hamstrings:
Biceps femoris
Semitendinosus
Semimembranosus

Iliotibial tract

Leg
Gastrocnemius

Soleus

Fibularis longus
Calcaneal (Achilles) tendon

different types of muscle, from the powerful muscles that move the arms and legs to the delicate muscles that open and close your eyelids.

to vital organs helps to hold those organs in place. A visual analogy is the Golden Gate Bridge in San Francisco, California, which is a classic example of a suspension bridge. Massive cables connect the main platform of the bridge to the metal structure that suspends it. These cables stabilize the platform to the frame much as the muscles stabilize the bones of the skeleton and the vital organs to the bones or other tissues.

The fifth and final critical function of the muscular system is the generation of heat. Muscle cells burn large amounts of **glucose**, a simple sugar that is the primary fuel for the cells of our bodies. The energy from using this glucose drives muscle movement, but it also generates heat. This is the reason why you become overheated if you undergo intense or prolonged exercise. Your skeletal muscles are generating more heat than the body needs to maintain its normal temperature. Under conditions of extreme cold, the voluntary muscles will undergo an involuntary process called **shivering**. The body triggers shivering if the temperature of the **trunk**, the central core of the body consisting of the chest and abdomen, begins to drop. The body will cause the skeletal muscles to undergo uncontrolled spasms, causing the body to shiver. This generates heat and helps to maintain the temperature of the body.

CONNECTIONS

As you see, the skeletomuscular system, a combination of the skeletal system and the muscular system, is a closely integrated system. We have seen that the skeletal system and the muscular system each have a number of critical functions. Some of these functions are absolutely dependent on each other, while others are more unique to the specific system. As we proceed through this book, we will learn more details about the components of the skeletomuscular system and will explore the common and the unique characteristics and functions of these components. Using the bones and muscles of your hands and arms, turn the page and begin your exploration of the "movers and shakers" of the human body.

2

Bones and Other Skeletal Components

TYPES OF BONES

With a total of 206 bones in the human body, you might ask if there is any way to group these bones together. Generally, scientists group or classify bones based on their shape. The four major classes of bones are long bones, short bones, flat bones, and irregular bones (Figure 2.1).

Long bones are much longer than they are wide. The central portion of a long bone, called the **shaft**, is surrounded by the **ends**. The shaft of a bone is also called the **diaphysis**, and the ends are called the **epiphyses** (singular is *epiphysis*). All the bones of the legs except the **kneecaps** and **ankle bones** are long bones, and all the bones of the arms except the **wrist bones** are long bones. The name "long bones" can be misleading. Many of the long bones, including those of the hands and feet, are actually quite small. The term *long* refers to their relative shape, not their size.

By contrast, **short bones**, such as those found in the wrist and ankle, are nearly as long as they are wide and thick. This gives the bones an almost cube-like shape, like the shape of dice. One special group of short bones is the **sesamoid bones**. These bones usually have one rounded end and a more pointed end. They are shaped similar to sesame seeds; thus, the name "sesamoid." The **patella**, or kneecap, is an example of a sesamoid bone.

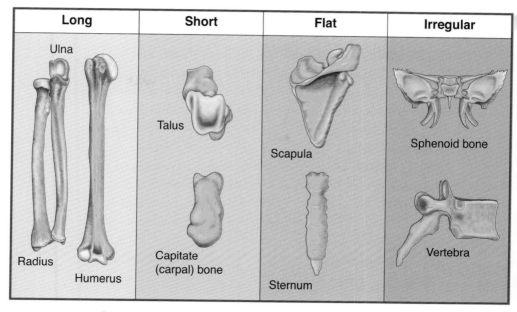

Long	Short	Flat	Irregular
Ulna Radius Humerus	Talus Capitate (carpal) bone	Scapula Sternum	Sphenoid bone Vertebra

Figure 2.1 The bones of the human body can be divided into four major categories: long, short, flat, and irregular. Each type of bone gets its name from its general shape, as seen in these diagrams.

The third group of bones is the **flat bones**. Flat bones tend to be wider than they are thick. Flat bones include the **sternum**, or breastbone; the **scapulae** (singular is *scapula*) or shoulder blades; the ribs, the series of bones that protect the chest cavity; and most of the **skull**, the bones that make up the head and jaw.

The final class of bones includes those that do not fit neatly into one of the other three categories. These are called **irregular bones**. Among the irregular bones are the **hip bones**; the **vertebrae** (singular is *vertebra*), or bones that form the **spinal column**; and the bones of the inner ear.

WHAT DO BONES DO?

Bones are the foundation on which the rest of the body is built. As a result, they are the first components that define our shape and form. In addition, there are a number of specific functions for the bones of the body, some of which may not be as obvious

23

as others. It is also important to remember that not all bones serve the same function. Each bone is specialized for its location and the job it must perform. Even bones on the right side of the body are slightly different from the same bone on the left side of the body, as they have mirror-image curvature rather than identical curvature.

The first basic function for the skeletal system and its bones is support. As noted previously, bones serve as a framework to which the other organs and tissues of the body are attached. If you have ever seen a contractor building a house, you may have noticed that the first thing to go up (after the foundation is laid) is the frame. It is the framework that supports the house, divides it into rooms, and to which the outside walls, the inside walls, the floor, the ceiling, and the electrical and plumbing components are attached. The frame of a house is very sturdy. If constructed properly, the house can withstand tremendous forces and remain intact, often with only cosmetic damage. So, too, the framework of your skeleton provides the strength for your body. In order to meet the demands of this function, the bones must be strong. The bones of the legs must hold the weight of the entire body, and when we run or jump, we increase the force on the leg bones many times over. The rib cage must hold the weight of the chest off of the lungs and heart so they can function. The bones of the arms, working with the muscles, allow us to pick up items much heavier than our arms themselves. These are only a few examples of how the bones provide support.

The second primary function of the bones is protection. Bones act as an armor of sorts. In this case, the armor is covered with a layer of skin and muscle, but certain bones are protective nonetheless. The bones of the cranium, a part of the skull, protect the soft tissue of the brain, the master control for the body. The vertebrae surround and protect the spinal cord, the main communication cable between the brain and the rest of the body. If either of these organs is damaged, the body will either cease to function or may be permanently impaired. The rib cage protects the lungs and heart from damage from outside the body, and the bones of the **pelvis** cradle the internal organs and, in pregnant women, the developing fetus.

The third function for the bones is movement. Although the muscles are critical for movement, the muscles must have a solid structure to work against. The ends of the skeletal muscles are attached to the bones. The bones then act as levers, magnifying the power of the muscles and allowing specific parts of the body to move. The muscles of the legs would not allow us to walk were they not attached to the bones of the pelvis, legs, and feet. Likewise, the muscles of our hands and arms would not make ordered and powerful movements were they not attached to the shoulders, arms, and hands. Even the process of breathing would not be possible were it not for the spinal column, the ribs, and the sternum. Both the muscles and the bones are essential for the graceful movements our bodies make.

The fourth function of the bones is the formation of blood cells. Blood cells are formed from special cells in the bone marrow, or soft center of many bones. In a process called **hematopoiesis**, these **bone marrow stem cells** give rise to all of the critical cells of the blood. If our bone marrow becomes defective, it can cause problems in the body. If too many white blood cells are made from bone marrow, a form of cancer called **leukemia** results. If not enough red blood cells are made, a condition called **anemia** results. Not surprisingly, the action of the bone marrow stem cells is carefully regulated in the body. Without enough red blood cells, called **erythrocytes**, we cannot get oxygen from the lungs to our tissues nor can we remove the carbon dioxide that results from metabolism in our tissues. Without enough white blood cells, or **leukocytes**, our immune system cannot protect us properly. The synthesis of blood cells is an often overlooked, but vital, function of the bones of the skeleton.

THE BONES WILL TELL— FACT AND FICTION!

In ancient times, fortune-tellers would take the bones of an animal (usually a sheep), mix them, and toss them out of a bag to form a pattern, in the manner of throwing dice. The fortune-tellers believed they could predict or "divine" the future based on the patterns that the bones formed, a practice called astragalomancy. Although there is no evidence that this practice was effective, a group of modern detectives has learned to discover facts about crimes based on the bones of the victims. **Forensic science** involves studying evidence from a crime scene to learn more about the crime and what happened to the victim. Certain forensic scientists are known as forensic anthropologists. They specialize in studying the bones of a victim to determine what they can about the victim and what became of her or him.

What kinds of things can a forensic anthropologist find out? Based on the bones alone, these scientists can determine the age, race, sex, and relative size of the victim. In some cases, they can accomplish this with only one or just a few of the 206 bones of the human body. Based on the mineral composition of the bones, the scientist can tell whether the victim died from certain types of poisons. The forensic anthropologist can use X-ray records to confirm whether a victim matches the X-rays of known missing persons (this is even possible using only dental X-rays). The bones can indicate if the person was a victim of repeated physical abuse, and can provide useful information about the diet of the victim. Finally, if a source of DNA is available from a suspected victim or her or his relatives, DNA can be extracted from the victim's bones and sequenced, allowing a definitive match of the bones to a particular person.

Analysis of bones to solve crimes has become a fascinating subject in detective fiction. If you are interested in this topic, you can read about the work of fictional scientists Dr. Kay Scarpetta in the books of author Patricia Cornwell, or of Dr. Temperance Brennan in the novels of Kathy Riechs.

The fifth function of our bones is to serve as a reservoir for minerals. Bones contain high concentrations of the elements calcium and phosphate. Both of these elements are essential to our bodies. When we do not get enough calcium and phosphate from our diet, the body will remove, or **leach**, the needed minerals from our bones. This causes the bones to weaken, which can lead to deformity or breakage. You may have heard that women need more calcium than men. This is generally true because women lose significant amounts of calcium during **menstruation**. This calcium must be replaced. As we age, our bodies become less efficient at incorporating calcium and phosphate into our bones, with the result that our bones become weaker and more brittle. Older individuals are often "stooped;" that is, their shoulders roll forward and their backs curve. This is due to loss of calcium and phosphate, which weakens the bones, allowing them to deform. Calcium is also essential for muscle strength, and as the muscles become weaker, the weight of the head tends to tilt the head and shoulders forward. Exercise is the best way to avoid stooping. It not only strengthens the muscles, but also promotes incorporation of calcium and phosphate into the bones. You may also have heard that older people are more likely to fracture their hips. The hips are the center point of where the weight of the body is concentrated. As the bones of the pelvis lose calcium and phosphorus, the bones become more brittle and are more prone to break. Hip fractures are particularly difficult to heal because of the constant stress that is applied to the pelvis. For this reason, hip fractures are very serious injuries for the elderly, and many older adults never fully recover from a broken or fractured hip.

The final function of the bones we will discuss is communication. The tiny bones of the inner ear transmit vibrations from the **tympanic membrane**, or eardrum, to other structures of the ear and ultimately along a nerve channel to the part of the brain that processes sound. Without these three tiny bones, we would not be able to hear. Likewise, our skeleton plays an important role in our ability to speak. A special type of cartilage, the **respiratory cartilage**, forms the larynx, or voice box, which allows us to generate the vibrations that eventually become sounds and words. We will discuss the role of soft skeletal tissues in Chapter 5.

TAKING A CLOSER LOOK AT BONE

A bone appears, at first glance, to be a solid structure, like a rock. But living bone is actually a complex network of channels and solid sections (Figure 2.2). If you were to take a thin section of bone and look at it under the microscope, you would see these channels. Each channel has two parts. The outer portion of the channel is a series of concentric rings that form the **osteon**. The osteon is shaped like a cylinder and runs parallel to the longest axis of the bone. The opening in the center of the osteon is called the **Haversian canal**. Through these Haversian canals travel blood vessels and nerves of the bone.

The layers that make up the osteon are well designed for strength. The layers of the concentric rings consist of long **collagen fibers** composed of tough connective tissue. These fibers are arranged in a helix, or spiral, shape; rather than traveling in a straight line, they curve around the central axis of the canal, like a spring. This spiral structure contributes to the strength of the osteon, but the structure goes one step further, in that each individual layer of the concentric rings spirals in the direction opposite the layers on either side of it. By alternating the direction of the collagen spirals, the osteon becomes extremely strong. A closer look at the bone section under the microscope reveals another group of channels that move away from the Haversian canals at right angles. These are the **Volkmann's (perforating) canals**. These canals provide for blood and nerves to enter the bone from the periosteum.

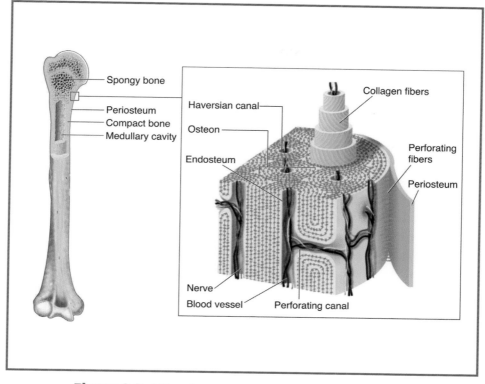

Figure 2.2 Although most people rightly think of their bones as extremely sturdy structures, bones are not, in fact, completely solid, but instead are composed of channels as well as solid sections. As seen in this diagram of a long bone, the layers and channels are designed to strengthen the bone while providing access for blood vessels and nerves.

The Periosteum

The **periosteum** is a double membrane that surrounds the outside of the bone. The prefix *peri* means "around," and the root *osteum* means "bone." The membranous coat of the periosteum consists of two layers. The tough fibrous outermost layer serves as a protective coating. The inner layer, called the **osteogenic layer**, is responsible for the growth and reshaping of bones. (*Osteo* means "bone," and *genic* means "to make or create.") Two basic types of cells are found within this layer: **osteoblasts,** or bone-building cells, and **osteoclasts**, or bone-destroying cells. An easy way to remember these cells is that "osteo**B**lasts **B**uild," while "osteo**C**lasts **C**runch" (destroy) bones. The periosteum is anchored to the bone itself by bits of collagen called **Sharpey's perforating fibers**.

The Endosteum

The long bones have a hollow core. This core is lined by another membrane, called the **endosteum**. The word means "inside the bone" (*endo* means "inside" and *osteum* means "bone;" thus, "inside the bone"). This membrane also lines the canals of the bone. Like the periosteum, the endosteum contains osteoblasts and osteoclasts, allowing bone to grow from the inside as well as the outside.

THE COMPOSITION OF BONE

Bone contains both organic and inorganic components. Organic components contain substantial amounts of carbon, hydrogen, and oxygen, while inorganic components are rich in minerals. The organic parts of bone include the osteoblasts and osteoclasts; **ground substance**, which consists of glycoproteins (**proteins** modified with sugars) and **proteoglycans** (sugars modified with **amino acids**); and collagen. The remainder of bone (about 65%) is composed of inorganic salts, mainly calcium phosphate. The organic components, particularly the collagen, account for the resilience of bone (its ability to resist breaking when stressed), while the inorganic components account for its hardness.

CONNECTIONS

In this chapter we have taken a closer look at bones. We have learned that they come in a variety of shapes and types, and that each bone contributes to the function of the part of the body where it is located. We have learned that bones play a number of roles in the body. Some of these roles, such as providing a framework and protection, are familiar to you, but other roles, such as the production of blood cells and inorganic ion storage, may be surprising. We have seen that the bone is more than a solid, rigid rod. It is a network of canals surrounded by layers of collagen and calcium phosphate. This network of canals is critical for bone growth and change. We have also seen that bone cells are specialized into those that build up the bone matrix and those that remove it. The action of these cell types makes bone dynamic and constantly changing. In the next chapter, we will see that the bones of the human body can be divided into two units, each with a specific role to play in our bodies.

3

The Axial Skeleton

THE GAME PLAN

Composed of 206 bones, the human skeleton appears to be a challenge to study. Fortunately, the human body has developed in such a way that there are **lines of symmetry**, imaginary lines that can be drawn through the skeleton so that the parts of the skeleton on either side of the line are mirror images, or near mirror images, of each other. The most important line of symmetry for describing the skeleton is the **medial line** (*medial* means "middle"). This is an imaginary line, also called the midline, that runs from the center of the skull, through the trunk of the body, and directly between the legs. It is a line from head to foot that divides the left side of the body from the right side of the body. Thus, the medial line is one that runs along the center **axis** of the body. Unless there has been a birth defect or an injury, your right arm is a mirror image of your left arm, and your right leg is a mirror image of your left leg.

The skeleton is divided, for ease of description, into the axial and appendicular skeletons. The **axial skeleton** includes the skull, vertebral (spinal) column, and the bony thorax, or cage of bone, consisting of the thoracic vertebrae, the ribs, and the sternum (breastbone), which surrounds the chest. While these bones vary greatly in their appearance, they all share one thing in common. The primary function of all the bones of the axial skeleton is to

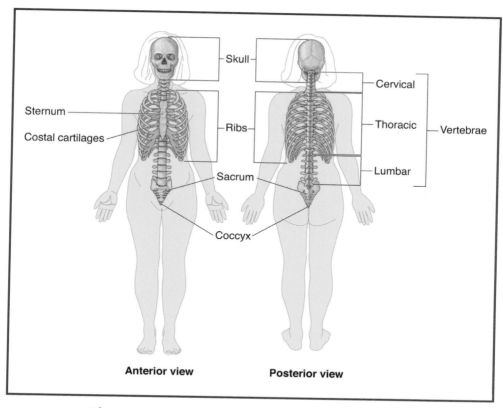

Figure 3.1 This diagram depicts the axial skeleton, including the skull, the spinal column, and the ribs. The primary function of the axial skeleton is to protect the central nervous system and the heart and lungs. Housed inside the sturdy series of bones, the brain and other vital organs are kept safe from trauma.

prevent damage to the delicate central nervous system or to the lungs and heart. The rest of the bones of the body belong to the **appendicular skeleton**. The word *appendicular* comes from the same root word as *appendix*, meaning "an appendage" or "something added on." The appendicular skeleton consists of those bones that are added on to the axial skeleton. As we will see later in this chapter, the appendicular skeleton includes the bones of the arms and legs, and the bones needed to attach these to the axial skeleton. In this chapter, we will explore the bones of the axial skeleton (Figure 3.1), and the bones of the appendicular skeleton will be the focus of Chapter 4. The goal of these chapters is not to memorize each bone of the body, but rather to understand how certain bones or groups of bones work together for a specific purpose. You can always refer to a textbook for the names and locations of different bones. We will focus on the form and function of bones and bone groups.

THE SKULL

The skull consists of the bones of the head (Figure 3.2). We can further divide these bones into two major groups. The **cranium** is made up of the bones that surround the brain. The brain is probably the most complex organ of the body. It is the "central processor" for every other organ in our body and is composed of soft tissues. Because of its central importance in body function, the brain must be carefully protected. The bones of the cranium serve as a "helm"—the section of a knight's armor that protects the head.

The cranium can be divided into two functional groups: the vault and the base. The word *vault* brings to mind a highly protected safe or chamber. This is an apt description of the **cranial vault**. It is used to store and protect the most valuable treasure of the body: the brain. The cranial vault makes up the top, sides, and back of the skull. The other section of the cranium is the **cranial base**, which consists of the bottom of the skull. When the vault and the base are joined, they create a chamber, the **cranial cavity**. It is within this highly secure cavity that the brain resides, protected on all sides from damage.

The cranial vault is made up of eight plate-like bones that curve out slightly in the middle. These bony plates meet at lines called sutures. When we are born, these sutures are not connected. This allows the bones of the cranial vault to move relative to each other. This is important when the baby is born because it gives the skull flexibility as it passes through

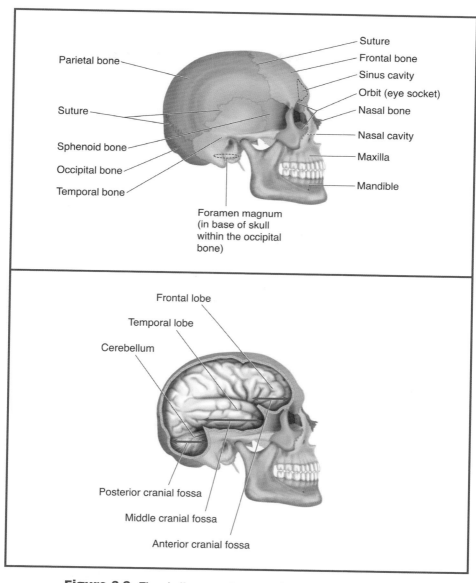

Figure 3.2 The skull, or cranium, consists of all the bones of the head, illustrated here. It has two major components: the vault and the base. The vault surrounds and protects the brain and consists of a series of flat bones that join together. The base consists of the point at which the spinal cord fuses to the brain and the lower jaw.

the birth canal. A newborn baby's skull is sometimes elongated or otherwise misshapen immediately after birth, but can be gently massaged and shaped to realign the bones of the cranial vault into their normal positions. Because the bones of the cranial vault will continue to grow for several years, the bones may not completely enclose the brain. You have probably been told that you must be very careful with the head of an infant because there are gaps or "soft spots" between the bones. A blow to one of these spots could result in permanent damage to the brain. The cranial base is divided by internal ridges that form three steps, or **fossae** (singular is *fossa*). These fossae hold the brain snugly in place.

The cranial chamber is not the only cavity, or hole, in the skull. There are cavities on either side of the skull for the middle and inner ears, and two **orbits** on either side of the front of the skull that house the eyes. Below the orbits is the nasal cavity, which is critical for our sense of smell and for respiration. There are also many other "openings" in the skull. These are known by a number of names, including **foramina**, or hole; **canals**, or narrow tubes or channels; and **fissures**, or crevasses or gorges. An opening in the base of the cranium, the **foramen magnum**, serves as the point of entry for the spinal cord, and there are small cavities in the skull that accommodate the **sinuses**, small air-filled pockets that connect to the respiratory system.

The bones of the face are also considered part of the skull. The **nasal bone** is found at the top of the nose where it joins the skull. If you take your thumb and forefinger, place them on either side of the tip of your nose, and wiggle gently, you will probably notice that there is considerable flexibility. This portion of the nose does not contain hard bone, but a softer flexible skeletal tissue called **cartilage**. Cartilage does not contain a high concentration of calcium phosphate. If you now gently slide your thumb and forefinger up the sides of the nose you will feel a slight ridge. If you wiggle your thumb and finger above this ridge you will notice that it is very rigid. This is the hard nasal bone. On either side of the nasal bone is the **maxilla**. The maxilla extends down and around from the nasal bone, encircling the nasal cavity, and serves as the anchor site for the upper row of teeth. Attached below the cranium, on the front side, is the **mandible**, or jawbone. If you move your jaw, you will see that it moves up and down, and wiggles side to side. The mandible joins to the cranium at two points on each side. These points are rounded and attached in such a way as to allow a wide range of mobility, the mobility that allows us to chew our food and form sounds for speech. The mandible is the strongest bone of the face. Starting at the chin, the mandible is relatively flat as you move backward on both sides. Somewhere in a line below the ear, the bone angles upward to ultimately come in close contact with the cranium. If you run your fingers along the jaw from the chin backward, you can probably feel this angle. The mandible is the bone into which the lower teeth are anchored. While we have a high degree of rotation around the contact points between the mandible and cranium, we cannot completely disconnect these contacts. Many snakes can completely disconnect their mandible from the cranium, allowing them to swallow prey that is actually wider than their mouth. If you see a snake shortly after feeding, you may see a large "lump" partway down the body. This lump, the ingested prey, illustrates how disconnecting the mandible assists in feeding.

THE VERTEBRAL COLUMN

The **vertebral column** (Figure 3.3) is a protective sheath that surrounds the **spinal cord**. You might think that the best protection would be provided by a structure that was a hollow cylinder, like a piece of pipe, but the vertebral column requires a high degree of flexibility. If you stand and bend at the waist, you will notice that the back does not remain rigid,

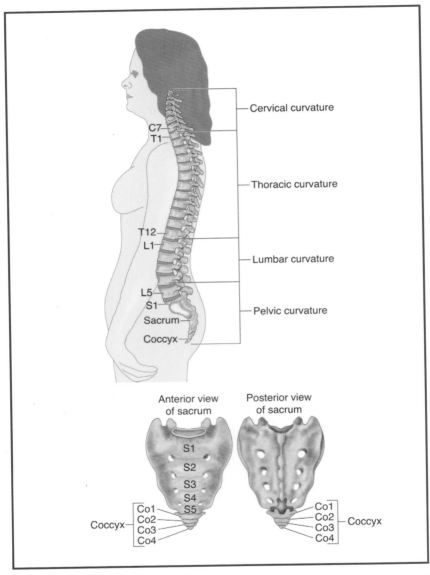

Cervical curvature

Thoracic curvature

Lumbar curvature

Pelvic curvature

Anterior view
of sacrum

Posterior view
of sacrum

Figure 3.3 The spinal column is composed of a series of irregular bones called vertebrae. The vertebrae are divided into groups based on changes in the direction that the spinal column curves. These groups are the cervical, thoracic, lumbar, and pelvic curvatures. The sacrum and coccyx are located on the end of the spinal column farthest from the skull. As we age, they form a spearhead-shaped mass of fused bone.

YOUR HEALTH—
SPINAL CORD INJURIES

The spinal cord serves as the main communication cable between the brain and the rest of the body. Different nerves enter the spinal cord at different positions along its length. Like other nervous tissues, the spinal column cannot repair itself. If you damage the "communications cable" at any place along the spinal cord, messages cannot get from the brain to the parts of the body that connect below the injured point. Thus, spinal cord injuries usually affect parts of the body on the lower, or "feet," end of the injured site.

The two most common types of spinal cord injuries result in either **paraplegia**, a condition that allows movement of the arms and upper body, but not the lower body, or **quadriplegia**, the paralysis of all four limbs of the body. Quadriplegia will be discussed later in this chapter. Paraplegia usually occurs when a person suffers a severe, forceful blow across the upper back, or between the last of the cervical vertebrae and the first thoracic vertebrae. This can happen to young people in automobile accidents or falls in which the person lands on his or her back. An example of the latter might be a fall from a horse while horseback riding or a backward fall while rock climbing. Damage to the spinal cord usually occurs somewhere within the thoracic vertebrae. This leaves the nerves of the head, arms, and upper chest unaffected, but will usually "disconnect" the nerves from the lower trunk and legs from the brain. The result is that the paraplegic generally is able to breathe normally and use the arms and the head and neck fully, but has no control over the legs and may have difficulty or a complete inability to control bladder and bowel functions. Paraplegics can usually live very productive lives, although they are most often confined to a wheelchair.

Many paraplegic accidents can be avoided if you are careful. Being a cautious driver and paying close attention to what you and others are doing will reduce the risk of paraplegia from an automobile accident. This is why it is so important to concentrate your total attention on driving when you are behind the wheel of a car. Avoid distractions such as cell phones and passengers who take your attention away from the road. Many automobile accidents can be prevented if you remember that the only focus for your attention when you are behind the wheel is driving—nothing else matters. Accidents such as those that are the result of falls from horseback riding or rock climbing are best avoided by being well trained in what you do. These athletic activities require special equipment and knowledge, and should not be attempted without proper preparation. The majority of paraplegic events that occur during sports happen to inexperienced participants.

but curves as you bend forward. Straightening again, notice that your hips and shoulders are probably aligned. However, if you place your hands on your shoulders and twist your shoulders you will see that the body twists beginning at the hips and continues twisting to the shoulders. Finally, if you straighten again and place your hands on your shoulders with your elbows pointing outward, you can drop your left elbow toward your feet and your right elbow will point upward. These movements show us that there is much flexibility in the vertebral column. It can bend, rotate, and shift from side to side.

This high degree of flexibility is a result of the many bones that make up the vertebral column. When we are first born, there are 33 different bones in the vertebral column. These bones are called vertebrae (singular is *vertebra*). Each vertebra is a circle of bone that forms a collar around a porous core (**centrum**, or "body"). Extending away from most of these collars are projections of bone called **pedicles** and **processes**. The processes circle around from each side and form a hole called the **vertebral foramen**, through which the spinal cord ascends to the brain. The pedicles and processes are locations for contact between vertebrae and for attachment of muscles and ligaments.

The Fused Vertebrae

You will recall that we mentioned that there are 33 separate vertebra in the spinal column of a newborn infant. As we age, some of these vertebrae change. The four vertebrae at the bottom of the spinal cord fuse together to form the **coccyx**, or tailbone. The vertebrae that make up the coccyx are smaller and simpler than other vertebrae and are free of processes and pedicles. While the coccyx is generally thought to be a **vestigial structure**—that is, a structure no longer important to human function—in other animals, it is the basis for the tail. The tail is often an important structure for balance and in some animals it is prehensile, meaning that it can be used for grasping and holding. In these animals, the bones of the tail are not fused. Occasionally, babies will be born with an unusually long coccyx, resulting in a short "tail" extending from the base of the spine. Physicians will usually remove this tail shortly after birth.

Above the coccyx are another five vertebrae that fuse as we mature. These form the **sacrum**. The sacrum is shaped like a spearhead that points downward. Not only are the collars fused, but the processes and pedicles are also reshaped and fused to form nearly flat panels with four holes in each side panel. The sacrum is the point for attachment of the hips.

The Unfused Vertebrae

Moving up from the coccyx to the top of the sacrum, the vertebrae gradually get larger. Above the sacrum are the vertebrae that are unfused in most individuals. As you move up toward the skull, these vertebrae gradually get smaller. We will discuss these vertebrae from top to bottom, moving from the skull back toward the sacrum.

The unfused vertebrae can be divided into three major groups. While the number of vertebrae in each group varies in 5% of the population, generally we have the same number of vertebrae in each group. How do we divide these vertebrae into groups? You may be surprised to learn that in a healthy person, the spinal cord is not straight, but curves. It actually changes direction at two points, dividing the vertebrae into three sections. The top section, always consisting of seven vertebrae, curves outward toward the back. These vertebrae are called the **cervical vertebrae** and the concave curve is called the **cervical curvature**. The 7 vertebrae, starting at the skull end, are called the C1 through C7 vertebrae. These are the smallest, lightest, and weakest of the vertebrae. Vertebrae C1 and C2 are different from the others, since they are the contact point with the skull and form the first vertebral joint.

The second section of the vertebrae usually consists of 12 vertebrae. These vertebrae are the **thoracic vertebrae**. These vertebrae curve in a convex manner; that is, they curve forward, toward the chest. As you move down these vertebrae from T1 at the top to T12 at the bottom, the vertebrae increase in size, weight, and strength. The thoracic vertebrae are the points where the ribs attach to the spinal column in the back of the body.

The third set of vertebrae, the **lumbar vertebrae**, usually consists of 5 vertebrae. These continue to enlarge as you move downward from L1 to L5. Like the cervical section, the curvature of the lumbar regions is concave, curving outward toward the back. The projections off of these vertebrae restrict rotation more than the cervical and thoracic vertebrae, providing a strong support for the base of the spinal cord.

One way to remember how many vertebrae form each section of the spine is to associate each with a typical mealtime. If you eat breakfast around 7 A.M., lunch around 12 noon, and an early dinner around 5 P.M., you can remember

QUADRIPLEGIA

Quadriplegia results in the paralysis of all four limbs. It is a more life-threatening condition than paraplegia because it affects the vital function of breathing. Many quadriplegics need mechanical assistance to breathe properly. Additionally, use of a wheelchair is more difficult because a quadriplegic does not have the use of the arms. In most cases, quadriplegia occurs when a severe blow is delivered directly to the skull. If the blow is delivered from the side, the sudden force may cause one or more cervical vertebrae to damage or sever the spinal cord. A blow from the front or the back can force the cervical vertebrae to pinch or cut the spinal cord. As a result, a quadriplegic may be left with control of only the head and perhaps some of the neck muscles. Mechanical wheelchairs are now available that can be directed with a stream of breath from the mouth, or by a machine that monitors the movement of the eyes.

One of the most common ways that young people can fall victim to quadriplegia is swimming accidents—more specifically, diving accidents. If a person dives into water that is too shallow, or is shallower than she or he anticipated, the person's head will hit the bottom of the pool or pond with tremendous force. This may immediately sever or damage the spinal cord, or it may fracture one of the cervical vertebrae. Diving accident victims must be carefully removed from the water, avoiding movement in the head or neck region, because a fractured vertebra may damage the spinal cord further even after the initial injury has occurred. This is why lifesavers move very cautiously when they remove a diving accident victim from the water. They will usually first carefully tie the victim onto a backboard to provide support for the neck and back, and to prevent movement of the head. These precautions reduce the likelihood of secondary injury. You can avoid these injuries by never diving in shallow water and never diving into water if you do not know the bottom conditions or its depth.

Why do diving accidents tend to cause quadriplegia more often than paraplegia? The most important factor is that the cervical vertebrae are smaller and weaker than the thoracic or lumbar vertebrae. As a result, they are more delicate and prone to damage, especially from a blow to the head.

One other common cause of quadriplegia (as well as paraplegia) in young people is motorcycle accidents. Unlike a car, a motorcycle provides little protection for its rider. Quadriplegic accidents can be significantly reduced, however, by simply wearing an approved motorcycle helmet. The helmet helps keep the head aligned with the body, which reduces strain on the cervical vertebrae and protects the brain from injury in the event of a crash. The most common cause of adult-acquired epilepsy (a condition that results in uncontrollable seizures of the body) is motorcycle accidents in which the rider was not wearing a helmet.

The purpose of this section is not to discourage you from swimming, driving, or riding a motorcycle. Rather, the goal is to emphasize the importance of proper training, practice, and use of the correct equipment, as well as undistracted focus on the task at hand. Paying close attention to possible risks will not only make for a more exciting and enjoyable experience, but may preserve your ability to live your life to the fullest in the future. You are your own best defense against spinal cord injuries. Though some cases are unavoidable, many can be prevented or minimized with proper care, training, and attention.

the cervical (7), thoracic (12), and lumbar (5) vertebrae. The fused vertebrae can be remembered by "f," for "fused." There are five fused vertebrae in the sacrum and four fused vertebrae in the coccyx.

The Bony Thorax

The **bony thorax** creates a cage of bone that surrounds the chest cavity (Figure 3.4). It consists of the sternum, or breastbone,

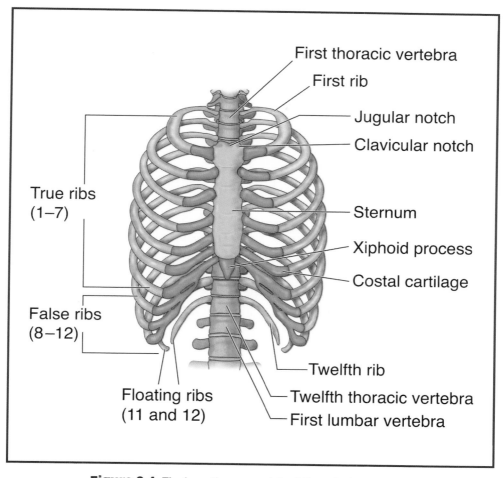

First thoracic vertebra

First rib

Jugular notch

Clavicular notch

True ribs
(1–7)

Sternum

Xiphoid process

Costal cartilage

False ribs
(8–12)

Twelfth rib

Floating ribs
(11 and 12)

Twelfth thoracic vertebra

First lumbar vertebra

Figure 3.4 The bony thorax consists of the spinal column, the ribs, and the bones of the breastplate. These form a strong cage that both protects the heart and lungs and helps drive the breathing process.

the 12 thoracic vertebrae, and the ribs. The sternum is a flat plate of bone in the center of the chest that results from the fusion of three bones. There is an indentation at the top of the sternum, called the **jugular notch**. If you gently allow your forefingers to travel down the front of your throat you should be able to feel this notch. From the notch, the sternum widens and then narrows again, forming a heart-shaped top to the bone. It extends downward in a dagger shape. The sternum finally narrows to a separate diamond-shaped flat bone called the **xiphoid process.**

The ribs are the major bony component of the bony thorax. The ribs are slightly flattened bones that curve from the spinal column, around either side of the chest, and end at the sternum in the front of the body. Starting at the top of the chest, the first seven ribs on each side are called the **true ribs**, since they connect directly to the sternum through short pieces of cartilage called **costal cartilage.** The next three ribs are called the **false ribs**, in that each of the three upper pairs is attached to the cartilage of the rib above rather than directly to the sternum. The remaining two ribs don't connect to the sternum at all (usually ribs 11 and 12) and are called **floating ribs.**

The ribs begin shorter at the top and get progressively longer from rib 1 to rib 7 (all true ribs), then decrease in length from ribs 8 to 12.

CONNECTIONS

The axial skeleton can be divided into three major sections. The skull consists of the bones that protect the brain and support the teeth. The primary functions of the skull are protection of the brain, eating, and speaking. The vertebral column consists of 33 bones, with the lower nine vertebrae fused into two structures: the coccyx and the sacrum. The other vertebrae are divided into three groups of unfused bones. Beginning at the skull, the first seven vertebrae form the cervical curve, the next twelve form the thoracic curve, and the final five bones form the lumbar curve, which rests above the sacrum. The primary function of the spinal column is protection of the spinal cord, while allowing reasonable rotation in the torso. The final group of bones in the axial skeleton forms the bony thorax, a cage of bone that surrounds the chest. This structure consists of 12 pairs of ribs, each pair of which attaches to the spinal column at a different thoracic vertebra. The top seven pairs of ribs are the true ribs and join directly to the sternum in the front of the body via costal cartilage. The next three pairs of ribs, the false ribs, only connect to the costal cartilage of the rib above. The remaining two pairs of ribs are known as floating ribs, since they do not connect to the sternum at all. The primary functions of the bony thorax are to protect the heart and lungs and to facilitate breathing.

4

The Appendicular Skeleton

AN ARM IS AN ARM, A LEG IS A LEG—
AN ARM IS A LEG?

The body's medial line, as discussed in Chapter 3, divides the body in half, so that the left arm is a mirror image of the right arm, and the left leg is a mirror image of the right leg. This symmetry is also observed in the bones that make up the arms and hands and the legs and feet. It is fair to say, even though there are slight differences, that an arm is an arm and a leg is a leg. If you look at an arm and compare it to a leg, they really don't look much alike. Or do they? If you compare the make up of the bones of the arms to those of the legs, you will find clear differences, no doubt, but you will also find many similarities. With a few notable exceptions, the composition of the bones of the arms and hands are very similar to those of the legs and feet. This is even clearer if you compare the forelimbs (our arm equivalents) of a four-legged animal, like a cat or dog, to the hindlimbs (our leg equivalents). The process of walking on two feet (bipedal) rather than four (tetrapedal) has allowed organisms that walk upright to select changes in the upper limbs that make them useful for grasping, holding, and maneuvering, thus accounting for the changes we see between our upper limbs and our lower ones. In this chapter, we will study the upper and lower limbs that are appended or "added to" to the axial skeleton. These include the shoulders, arms, hands, and fingers, and the hips, legs, feet and toes (Figure 4.1).

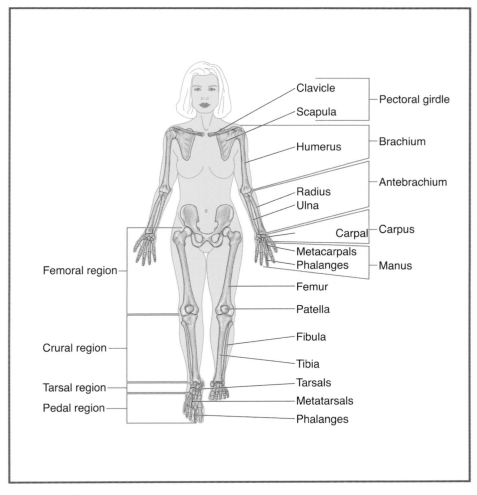

Figure 4.1 The appendicular skeleton consists of an upper series that contains the pectoral (shoulder) girdle and the arms and hands, and a lower series that contains the pelvic (hip) girdle, the legs, and the feet.

THE PECTORAL GIRDLE

The arms are connected to the axial skeleton at the upper region of the thoracic cavity by a series of bones collectively called a **pectoral girdle**. "Girdle" may seem like an odd name for this structure, but remember, many words have different meanings. While you may think of a girdle as an undergarment that helps to keep the stomach flat, a girdle is also a device that straddles a central axis. In this case, the pectoral girdle straddles the spinal column and upper portion of the thoracic cage. It works somewhat like a coat hanger to which the arm bones are attached. It consists of two bones on each side: the clavicle, and the scapula. The innermost bone is the clavicle, also called the **collarbone**—a long, thin bone that extends across the shoulders. Because this bone is so thin, it is easily broken. You may know someone who has suffered from a broken collarbone. People who lift weights or who do work that requires much upper body strength will actually have a thicker collarbone than those who do not do these activities. Not surprisingly, these people are also less likely to suffer breaks or fractures of their collarbone.

The collarbone attaches on one end to the sternum and on its other end to a **scapula** (plural is *scapulae*), or shoulder blade (Figure 4.2). The scapula is a triangular-shaped flat bone with the longer sides of the triangle angling down the back and the shorter side parallel with the top of the shoulder. Each scapula has three surfaces, called **fossae** (singular is *fossa*). Each fossa serves as an attachment point for a number of muscles that not only move the scapula, but also hold it in place. Three important regions on each scapula are the **acromion**, the outermost and upward point of the shoulder where the collarbone attaches to the scapula; the **coracoid process**, a finger-like piece of bone that serves as the attachment point for the biceps and other muscles of the arms; and the **glenoid cavity,** where the upper end of the **humerus**, or upper arm bone, attaches.

One of the most interesting facts about the pectoral girdle and its bones is that all of them join with loosely fitting joints that allow a wide range of motion. This allows us to have the mobility we need for our upper limbs and provides much more mobility and range of movement than we find for the lower limbs. Although this freedom of movement is important to humans, it is essential for other primates, such as monkeys,

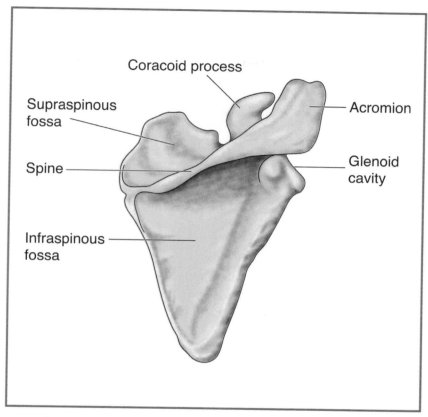

Figure 4.2 The scapula, or shoulder blade (illustrated here), takes the shape of an irregular spearhead. It is composed of the coracoid process, supraspinous fossa, spine, and infraspinous fossa.

that must have very wide-ranging movement in their upper bodies to allow them to live in the tree canopy and swing from branch to branch. We will discuss these joints in more detail in Chapter 5.

The Upper Limb

Each upper limb, or arm, in the human body consists of 30 bones. We can divide these bones into four sections to assist in learning their locations and functions (Figure 4.3). The

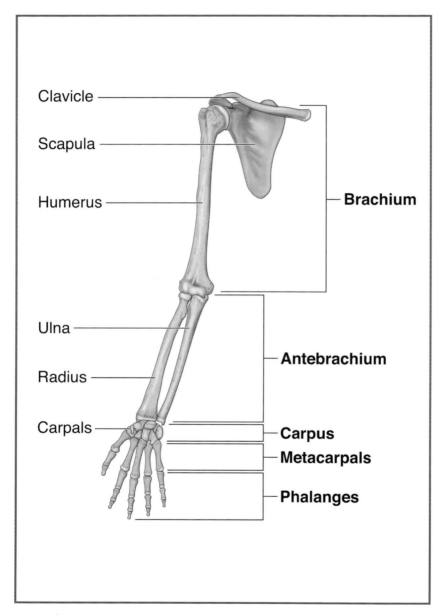

Figure 4.3 In this illustration of the upper right appendicular skeleton, the simpler structures of the upper arm bones are seen leading in to the complex workings of the wrist and hand bones.

brachium, also known as the upper arm, extends from the shoulder to the elbow. Each upper arm consists of a single long bone, the humerus. The **antebrachium** (*ante* means "before"), also known as the forearm, extends from the elbow to the wrist. There are two bones in the antebrachium: the **radius** and the **ulna**. The ulna is the slightly larger of the two bones.

The third region of the upper limb is the **carpus**, also known as the wrist. The carpus consists of eight small bones, the **carpal bones**, arranged in a cluster that roughly forms two rows. While the humerus, radius, and ulna are all long bones, the carpal bones are short bones. The fourth region of the upper limb is the **manus**, or hand. The 19 bones that make up each hand can be divided into two groups. The five bones of the palm constitute the **metacarpals** (*meta* means "situated beyond"). You can remember that these are between the fingers and the wrists because they are the metacarpals, an extension of the carpal bones of the wrist. Thus, the metacarpals are situated beyond the carpals. The remaining 14 bones of the hand are the **phalanges** (singular is *phalanx*), the bones of the fingers and thumb.

The Humerus

The humerus, the bone of the upper arm, has two very different ends (Figure 4.4). The end of the humerus that connects to the shoulder is a hemisphere, or half-sphere, and is called the "head." This hemispherical head joins the scapula at the glenoid process. The other end of the humerus has two smooth bumps or knobs called **condyles**. One of these, the more rounded, is the **capitulum**, which makes contact with the radius. The other condyle, the **trochlea**, is shaped more like a pulley and makes contact with the ulna. The humerus flares out just above these condyles, forming two bony complexes called the **epicondyles**. One of these, the **medial epicondyle**, is called the "funny bone" because it houses the **ulnar nerve**. A

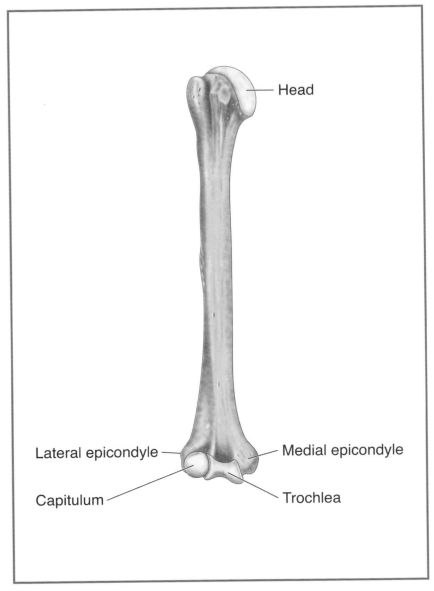

Figure 4.4 The humerus is the bone of the upper arm and is the largest bone of the upper appendicular skeleton. Each end has a slightly different structure. These varying shapes help create joints that allow the arm to move in different directions.

sharp tap at just the right spot stimulates this nerve, causing an intense tingling.

The Radius and Ulna

The end of the radius that attaches to the humerus is shaped like a disk and allows free rotation at this joint. This allows you to twist your palm upward or downward. The other end of the radius has three features. The **styloid process** of the radius is a projection of bone close to the thumb, the **articular facets** are indentations that interface with bones of the wrist, and the **ulnar notch** is a groove where the ulna makes contact with the radius (Figure 4.5).

The ulna attaches to the trochlea of the humerus through a cup-shaped pocket called the **trochlear notch**. This joint functions like the hinge of a door, allowing the forearm to move at a right angle relative to the humerus. The other end of the ulna connects to the wrist through its own styloid process. If you place the thumb and forefinger of one hand at the sides of the wrist joint of the other and slide the thumb and finger up and down the wrist, the bumps you will feel are the styloid processes of the radius and ulna.

The radius and ulna are connected to each other along their length by a special **ligament** called the **interosseous membrane**. This ligament allows the two bones to move over each other. This happens when you twist your wrist. Try it. Notice that not only does the wrist rotate, but also the entire forearm rotates. By having two bones rather than one in the forearm, the body can rotate one relative to the other, giving greater flexibility to the forearm and wrist.

The Carpal Bones

The eight bones that make up the wrist are arranged in two rows of four bones each (Figure 4.6). These are short bones for the most part. The first row of bones, closest to the elbow, contains the **scaphoid** (boat-shaped), **lunate** (moon-shaped),

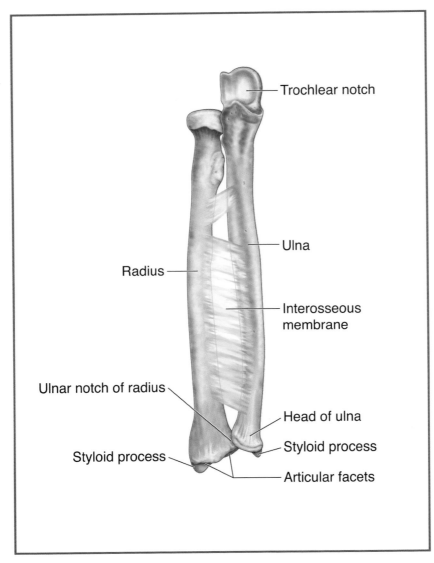

Figure 4.5 The radius and ulna are the bones of the forearm. The radius is the larger and stronger of the two, while the ulna serves as a strut, or stabilizing bone, for the radius. The radius and ulna connect to the humerus via the trochlear notch and the carpal bones of the wrist via the styloid process, articular facets, and ulnar notch.

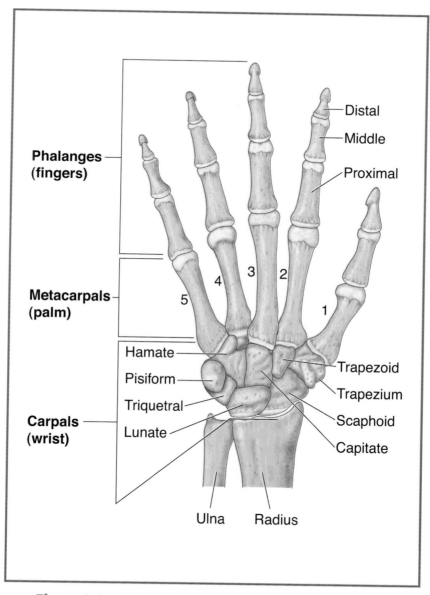

Figure 4.6 The hand and wrist consists of a complex interweaving of extremely specialized bones. Seen here are the bones of the left wrist and hand, including the phalanges, the metacarpals, and the carpals.

triquetrum (triangle-shaped), and pisiform (pea-shaped) bones. The pisiform bone is unique among the carpal bones in that it is not a short bone but a sesamoid bone. The four bones that make up the second row, the row closest to the hand, are the trapezium, an almost circular bone; the trapezoid, a four-sided bone with different angles; the capitate, the most rectangular of the bones in this row; and the hamate, the bone at the base of the little finger. A little hook of bone on the finger end of the hamate, called the hamulus, is a characteristic of and useful landmark to identify this bone.

The Metacarpals and Phalanges

Metacarpals and phalanges are examples of long bones. The five bones of metacarpals are numbered using Roman numerals, beginning with the thumb (I) and ending with the little finger (V). In each case, the end of the bone closest to the wrist is called the base, the shaft of each metacarpal is called the body, and the end closest to the fingers is called the head. You can see the heads of your metacarpals when you make a fist. The knuckles that bulge out at the base of each finger are the heads of the metacarpals. If you look at an X-ray of your hand, the metacarpals look like extensions of the fingers because they form straight lines with the bones of the fingers, or phalanges.

There are a total of 14 phalanges in the hand: The thumb has two phalanges (proximal and distal) and each of the four fingers, or digits, has three phalanges (proximal, middle, and distal). To demonstrate the difference, make a fist with your hand. The thumb bends in two sections, while each of the fingers bends in three sections. Each section is a phalanx. Instead of giving each phalanx a different name, we distinguish them by their position. Just as we begin numbering the metacarpals with Roman numeral I for the base of the thumb, the phalanges of the thumb are also designated with the numeral I. We also distinguish the phalanges by their relative closeness to the center of the body. The phalanx of the thumb closest to the metacarpal, and therefore closest to the center of the body, is called the **phalanx I proximal**. (*Proximal* means "closest to.") The other phalanx of the thumb would be **phalanx I distal**. (*Distal* means "farthest from" or "most distant to.") Moving to the index finger, the phalanges would be **phalanx II proximal, phalanx II middle**, and **phalanx II distal**. Based on this pattern, what would you name the three phalanges of the ring finger?

THE PELVIC GIRDLE AND APPENDAGES
The Pelvic Girdle

The pelvic girdle (Figure 4.7) of an adult differs from that of an infant or young child. This is due to the **fusion**, or joining together, of the bones of the pelvic girdle as we get older. The two fused sections of the spinal column, the sacrum and the coccyx, form the centerline of the pelvic girdle. On either side is the right or the left **os coxae** (singular). When we speak of both of these bones together we refer to them as the **ossa coxae**. Collectively, these four bone complexes form a fused product that is shaped somewhat like a very shallow bowl standing on its side, with the lower rim slightly more forward (toward the front of the body) than the upper rim. The pelvic girdle serves a number of important functions. Among these, it concentrates and balances the weight of the trunk over the legs, and cradles and protects the urinary tract, the reproductive tract, and parts of the digestive tract.

The os coxae is actually created from the fusion of three bones that are separate in infants and young children. These three bones are the **ilium**, the **ischium**, and the **pubis**. The ilium is largest and is toward the head. The ischium is toward the back of the body at the bottom of the os coxae, and the pubis is toward the front of the ischium. The pelvic girdle of a woman is generally wider and shallower than that of a man, a difference that is essential for the support of the developing fetus during pregnancy. This difference in the pelvic girdle

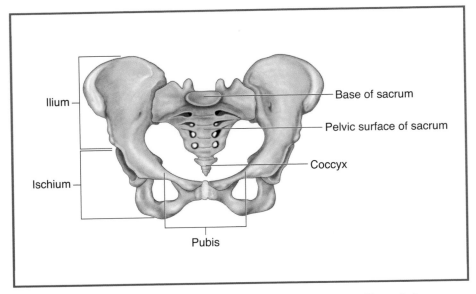

Figure 4.7 The bones of the pelvic girdle, or hips, are illustrated here. The pelvic girdle consists of the illum, ischium, pubis, and coccyx. Through its connections to the bones of the lower body and torso, the pelvic girdle allows you to twist, turn, and swing your hips.

allows forensic pathologists to determine the sex of a victim when only the skeleton remains. It also explains why women generally have wider hips, relative to the rest of the body, as compared to men.

The Lower Limbs

Just as we were able to divide the upper limb into four sections, similarities between the bones of arms and legs allow us to do the same for the lower limbs. As we found for the upper limbs, each of the lower limbs consists of 30 bones. The four regions of the lower limb are the **femoral region**, or thigh, which extends from the hip to the kneecap; the **crural region**, or lower leg, which extends from the knee to just above the ankle; the **tarsal region**, or ankle; and the **pedal region**, or foot and toes.

The Femoral Region

The longest bone of the body is the **femur**, the bone found in each femoral region of the body. Not surprisingly, it is similar in shape to the humerus of the upper arm. The femur attaches to the pelvic girdle through a ball-and-socket joint. The ball is on the end of the femur closest to the hip, and the socket is created at the intersection of the ilium, the ischium, and the pubis. The femur has a long shaft that extends down the thigh, and the distal end of the femur, like the humerus, ends in two condyles.

Although not precisely a part of the femoral region, the patella, or kneecap, is usually discussed along with the femur. The patella is a sesamoid bone that is triangular in shape. It forms within the tendon of the knee, the joint between the upper and lower leg, only after we begin walking. The patella is important in both protecting and strengthening the knee.

The Crural Region

Just as there are two main bones in the forearm, its leg equivalent, the lower leg, also consists of two long bones. These are the **tibia** and the **fibula**. The tibia is the larger and stronger of these two bones. It is critical for weight bearing. The fibula is a thinner bone that serves as a lateral strut. A strut is used to stabilize a structure, and the fibula stabilizes the ankle. The fibula is not designed to bear weight and would snap if the weight of the leg were not directed to the tibia.

The Tarsal Region

The bones of the ankle (Figure 4.8) are similar to those of the wrist in that they are made up of short bones that can be clustered in proximal and distal groups. They differ considerably in appearance and size, however, because the ankle must support the weight of the body while the wrists are not required to do this on a regular basis. The largest bone of the

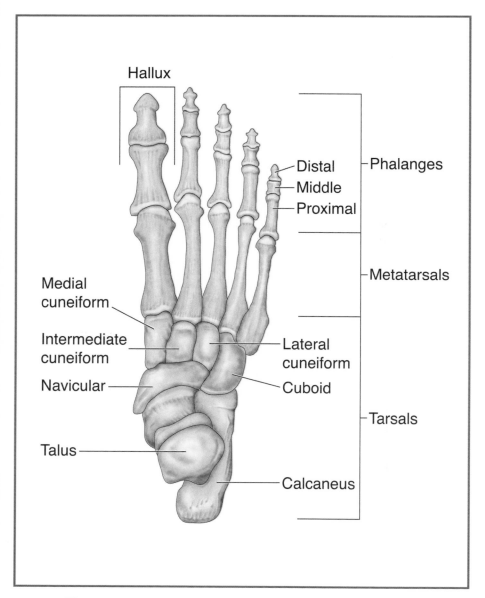

Figure 4.8 The skeleton of the foot is similar to that of the hand. The bones of the right ankle and foot are shown here. Like the wrist, the foot contains three groups of bones: the phalanges, the metatarsals, and the tarsals.

ankle is the **calcaneus**, or heel bone. The Achilles tendon joins the calf muscles to the calcaneus. If this tendon becomes cut or damaged, we can no longer bear weight on our leg, rendering the leg essentially useless. Adjacent to the calcaneus bone is the **talus**. The third bone in the proximal row is the **navicular**.

The distal row of bones in the ankle consists of four bones. The largest of these is the **cuboid**. The others are the lateral, intermediate, and medial **cuneiforms**.

The Metatarsals and Phalanges

The bones of the soles of the feet are metatarsals (Figure 4.8). Like the metacarpals of the hands, each foot contains five metatarsals. These are numbered with Roman numerals as well, beginning with the metatarsal of the big toe and ending with the little toe. Metatarsals and phalanges are all long bones. Metatarsals I, II, and III connect respectively with the medial, intermediate, and lateral cuneiforms. Metatarsals IV and V both connect with the cuboid. The big toe, also known as the **hallux**, only contains a proximal and a distal phalanx. The other four toes of each foot contain a proximal, intermediate (middle), and distal phalanx. The naming and numbering of the phalanges of the feet are identical to those of the hands.

The bones of the feet and toes are positioned so that they form arches on the sole side of the foot. The arches are held together by ligaments and are important in maintaining our balance. Too much weight or constant use of the feet will stretch these ligaments and will cause a condition known as fallen arches or flat feet, which makes prolonged walking and standing uncomfortable and difficult. Good-quality shoes have arch supports in them to take the stress off the ligaments of the feet. By wearing good-quality shoes, constant walking, running, and standing become less damaging and uncomfortable.

FOR YOUR HEALTH—BUNIONS

You may have heard of people having bunions, but do you know what they are? Although you probably think of these as problems only in old people, bunions can happen at any age, though they tend to get worse as we get older. A bunion is created when one of the bones of the feet shifts out of alignment, producing a protruding "bump." These become irritated and can make walking difficult.

Bunions can be caused by a lot of things. They may result from an accident that damages the bones of the feet. They may be caused by arthritis, in which swelling of the soft tissue of joints forces the bones out of alignment. A common source of bunions is wearing shoes that do not fit properly—especially narrow or pointed shoes. Imagine that your feet are a bag of marbles. The skin and muscles make up the bag, while the marbles are the bones. As long as you do not squeeze the bag of marbles, the bag will retain its shape. However, if you put pressure on one side of the bag, or squeeze it in the middle, some of the marbles will stick out. Now imagine that you coat the marbles with superglue before putting them in the bag. If you hold the bag the same way for a length of time, the marbles will create a pattern and hold it. If this lasts long enough, the glue will set and the marbles will remain in their position even after you stop squeezing the bag. A bad pair of shoes can squeeze the bones of your feet just like the bag of marbles. If you wear the shoes long enough, the bones remain out of alignment.

How are bunions treated? Usually, a patient goes to a specialist called a podiatrist (a doctor who treats problems of the feet). If the bunions are bad enough, the podiatrist will surgically expose the bunion to cut or shave off the protruding bone. Once the surgical site heals, the bunion no longer protrudes and does not irritate the foot anymore. However, if you continue to wear "bad" shoes, the bones of the feet will keep shifting and it may become necessary to repeat the surgery. When it comes to your feet, support and comfort are much more important than fashion. Wear "sensible shoes" and you and your feet will benefit.

CONNECTIONS

The appendicular skeleton consists of the pectoral girdle and upper limbs (arms, wrists, and hands), and pelvic girdle and lower limbs (legs, ankles, and feet). The major bones of the arms and legs, and of the hands and feet, consist primarily of long bones. The bones of the wrists and ankles are primarily short bones. Each upper limb and each lower limb is composed of 30 bones. There are many similarities in the composition of the bones of the upper limbs and the lower limbs; however, the demand of weight bearing has created noticeable differences in the individual bones of the upper and lower limbs.

5

Joints and Soft Tissues of the Skeleton

THE ROLE OF THE JOINTS AND SOFT TISSUES OF THE SKELETON

We have already learned that the skeletal system is made up of bones and that these bones come together, or **articulate**, at joints. Whenever two surfaces move against each other, part of the energy expended is converted to friction, the force that works in opposition to the direction of movement. Friction can be reduced if the surfaces are lubricated. The joints of the body offer their own lubrication. This is often accomplished through soft, cushioning tissues. In this chapter, we will learn how joints allow and affect the movement of bones, and we will discover the important role that soft tissues play in protecting joints and improving their efficiency. The study of joints and their functions is called **arthrology**.

Classification of Joints by Freedom of Movement

There are a number of ways to classify joints. One way is to consider the degree of freedom of movement at the joint. Joints like the elbow and shoulder, which have high degrees of freedom of movement, are **diarthroses** (singular is *diarthrosis*). Joints like those of the fingers and those between vertebrae, which have a more limited range of movement, are known as a **amphiarthroses** (singular is *amphiarthrosis*). Finally, some joints have very limited, if any, freedom of movement. Joints like the sutures between the plates of the cranium and the fused vertebrae of the lower back are known as **synarthroses** (singular is *synarthrosis*).

Classification of Joints by How They Join Adjacent Bones

Joints can also be classified into four major categories by how they adjoin adjacent bones. These categories are fibrous joints, cartilaginous joints, bony joints, and synovial joints.

Fibrous Joints

In **fibrous joints**, the collagen fibers from one bone extend and integrate into the adjacent bone. Thus, the two bones are physically joined by collagen fibers. There are three basic types of fibrous joints: sutures, gomphoses, and syndesmoses.

Sutures are fibrous joints that closely bind the adjacent bones and do not allow the bones to move. These are only found in the joints in the skull. There are three types of sutures. **Serrate sutures** are connections by wavy lines, increasing the total surface of contact, and thereby making the suture strong. **Lap sutures** are those where the bones have beveled edges that overlap. Finally, **plane sutures** occur where two bones form straight, nonoverlapping connections.

The joints that hold teeth into their sockets are called **gomphoses**. In this case, a strong band of connective tissue, the **periodontal ligament**, a structure made of collagen fibers, holds the tooth firmly to the jaw, but allows a little "play" to tolerate the mechanical stress of chewing food.

The final group of fibrous joints, the **syndesmoses**, consists of joints where two bones are joined only by a ligament. These are the most flexible of the fibrous joints. The interosseous membrane, the fibrous membrane that binds the radius to the ulna, and the tibia to the fibula, is an example of such a joint.

Cartilaginous Joints

When two bones are joined together by cartilage, the result-ing joint is called a **cartilaginous joint** (Figure 5.1). The two types of cartilaginous joints are **synchondroses**, which are joined by **hyaline cartilage**, and **symphases**, which are joined by **fibrocartilage**. An example of a synchondrosis is the joining of the ribs to the sternum in the thoracic cage. The fusing of vertebrae to one another is an example of a symphasis. In this case, each vertebra is covered with hya-line cartilage, but collagen fiber bundles connect the dif-ferent vertebrae together. Each connection has limited movement, but, collectively, these intervertebral discs make the spine flexible.

Bony Joints

Bony joints, or **synostoses**, result when two bones that were previously independent have fused together. This is often a normal process of aging. The process involves the gradual accumulation of calcium in fibrous or cartilagi-nous tissue through the process of **ossification**, the natural process of bone formation. This occurs between the epiphyses and diaphyses of the long bones as we mature. Synostoses also occur at other junctions between bones, including the jaw, and the pubic arch, a curvature of the pelvic girdle.

Synovial Joints

A **synovial joint** (Figure 5.2) is a freely movable joint in which the bones at the joint are separated by a lubricating, cushion-ing liquid called **synovial fluid**. Joints that do a great deal of work, including the elbows and knees, the hips, and the jaw, are often synovial joints. The synovial fluid does not freely move in and out of these joints. A special fibrous capsule, the **joint capsule**, which creates a sac-like pad filled with thick synovial fluid, surrounds the joint. The ends of the bones at

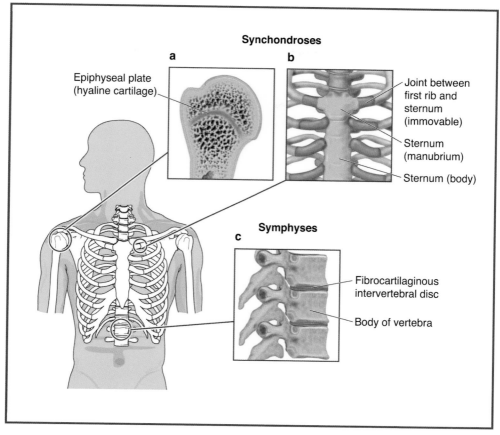

Figure 5.1 Cartilaginous joints are formed when two bones are joined by cartilage. Cartilaginous joints include synchondroses (e.g., the joint between the ribs and sternum) and symphases (e.g., the joints between vertebrae).

the joint are also coated with a fibrous membrane made of hyaline cartilage and are contained within the joint capsule. The thick synovial fluid continually bathes the joint, making the motion of the two bones relative to each other basically free of friction. In addition to lubrication, the synovial fluid acts as a shock absorber to reduce the effects of pressure on the joint.

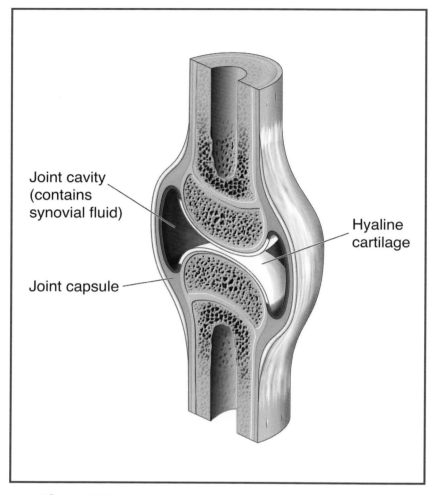

Figure 5.2 This is a cross-section of a synovial joint, in which the bones are held together by synovial fluid that helps cushion them from impact. Examples of synovial joints include ball-and-socket, hinge, pivot, gliding, saddle, and condyloid joints.

Types of Synovial Joints Synovial joints can be characterized by their structure (Figure 5.3). The **ball-and-socket joint** consists of one bone with a rounded end and the other with a cup to receive it. These joints are similar in design to a

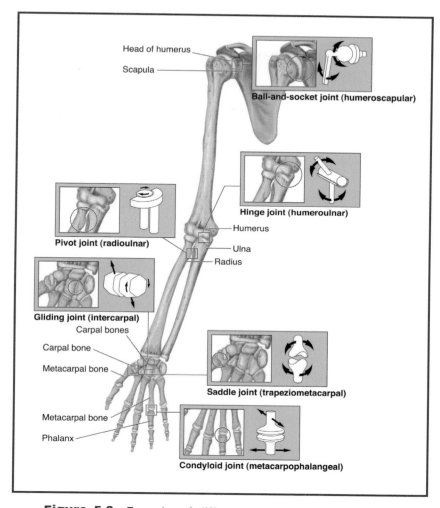

Figure 5.3   Examples of different types of synovial joints are illustrated here. Ball-and-socket joints allow bones to move in many directions, while hinge joints, gliding joints, and pivot joints allow bones to move in only one direction. Saddle and condyloid joints allow bones to move in two directions.

computer joystick or mouse ball, in that they have a high degree of freedom of rotation and make the joint very versatile in its range of movement. The shoulder joint and the hip joint are the two examples of ball-and-socket joints in the human body. These joints move in many directions and, as a result, they are called **multiaxial joints.**

A **biaxial joint** lets the bones move in two planes relative to each other. There are two main types of biaxial joints. The **saddle joint**, found only in the base of the thumb, allows the thumb to move up and down as well as toward and away from the other fingers. Because of the range of movement at this joint, the thumb can be brought over to contact the tips of the other fingers. This feature, called an opposable thumb, is only found in primates (humans, apes, and monkeys). A second type of biaxial joint is found at the base of the fingers, where the metacarpals and phalanges meet. These joints, which allow the fingers to move back and forth as well as up and down, are called **condyloid joints.**

The third general group of synovial joints consists of the **monaxial joints.** These joints allow movement in only one plane or direction. **Hinge joints,** like those found in the elbow, the knee, and the fingers and toes, are all monaxial because they have only one range of motion. Another example of a monaxial joint is the **gliding joint.** In this case, the ends of the bones slide over each other, allowing for limited movement in one direction. The final example of a monaxial joint is the **pivot joint.** In these joints, a projection from one bone fits into a ring of ligament on the other. The upper two joints of the spinal column are examples of pivot joints. One allows you to move your head side to side, while the second joint allows you to move it forward and back. We also find pivot joints where the ulna and radius meet in the forearm, and where the tibia and fibula overlap in the lower leg. These joints allow you to roll one bone relative to the other. In the arm, this results in being able to turn the palm up or down, and in the leg it allows the foot to angle in or out.

SPARE PARTS

Like any mechanical part, the joints and bones of the skeletal system can wear out. In some cases, a joint or bone may be injured beyond repair due to an accident. In other cases, disease may cause damage to bones or joints. Finally, as medicine has become more effective, we have begun to live longer. That means that some of our joints may wear out before we do. Orthopedic surgery allows specialists to replace damaged bones and joints. In some cases, bone transplants can be performed. Generally, this involves taking a healthy bone from an organ donor and using it to replace the damaged bone in another person. This technique is growing in popularity, but as with all organ donations, we need more people to volunteer to be organ donors. Your bones don't help you after you die, but they can be recycled and may give another person a new chance for life or for better quality of life. Being an organ donor is a wonderful way to help people in need. If you are not a designated organ donor, you may want to discuss becoming one with your parent or guardian. As with blood donation, most states require that you be 18 years of age or have parental approval, be in generally good health, and have a health history that is compatible with organ/blood donation.

In transplant cases that don't use donated bones, artificial joints or bones are used to replace those that are damaged or worn out. Knee and hip replacement surgeries are common now, particularly among older individuals. The joints used are often made of stainless steel, a material the body considers "neutral" and generally will not reject. Replacement of a damaged joint with an artificial one may restore mobility to the injured patient. Recently, scientists have begun to explore the possibility of using strong, high-impact plastics as alternatives to stainless steel joints. These materials are lighter and may lead to faster healing.

Compare your body to a car. When a part of the car wears out, you replace it with a new or used one. The same thing can be done when your bones or joints wear out. Replacement of these worn-out parts restores the damaged area and leaves you ready to rev your engine and go!

Though there are many other things we could discuss about the movement of joints, we will end our discussion on joint movement here. We will conclude the chapter by describing the different types of soft tissues associated with the skeleton.

Classification of Soft Tissues of the Skeletal System

There are three basic types of soft tissues associated with the skeletomuscular system. One type, which was previously described, provides padding to the joint. These sac-like pads contain synovial fluid and are similar to shock absorbers. A pad that is positioned between bones is known as a **meniscus**. We find these fluid-filled pads in the jaw and knees, among other places. Sacs of synovial fluid that are positioned between muscles, or where tendons pass over bone, are called **bursa** (singular is *bursae*). Specialized bursae that wrap around tendons are called **tendon sheaths**.

The second basic soft tissue of the body is the **tendon**. A tendon is a sheet or strip of tough collagen-containing connective tissue that are used to attach muscles to bones. The third type of soft tissue is the ligament. Ligaments are similar in structure and function to tendons, except they are used attach bones to bones. Tendons are important for the interaction of muscles and bones, and ligaments are important in the stable association and working of a joint.

CONNECTIONS

In this chapter, we have learned that the skeleton is more than just bones. Bones articulate with each other in different ways, forming joints that define the function and range of motion of a particular junction of bone. We have learned that joints can be classified by their mechanical design or by their range of motion and that these two methods of classification often overlap. We have also learned that soft tissues of the skeletal system—synovial pads, tendons, and ligaments—are essential for proper functioning of the skeletomuscular system.

6

How Bones Grow, Shrink, and Repair

WHAT IS BONE?

In the previous chapters, we have discussed the four types of bones—long, short, flat, and irregular—and the arrangement of bones in the human skeleton. In this chapter, we consider the characteristics of bone and how bones grow, shrink, and repair. The technical name for bone is **osseous tissue**. Osseous tissue is a specialized **connective tissue**, a tissue that functions mainly to bind and support other cells, that is hardened by the addition of calcium phosphate through a process known as **mineralization**. While osseous tissue gives bone its characteristics of strength and hardness, there are other types of tissues in bone. Osseous tissue containing lots of calcium phosphate is called **compact bone**, or dense bone, and is organized in cylinders that run parallel with the longest dimension of the bone. Compact bone is common throughout the shaft of weight-bearing bones, such as the long bones, and on the surfaces of other types of bones. These cylinders are a strong structure and contribute to the strength of the bone. Inside the cylinders is a space called the **medullary cavity**, which contains bone marrow. This complex compact bone and medullary cavity is most often found in the shaft, or central section, of a long bone. The ends of the long bone contain a different osseous tissue called **spongy bone**, bone with many spaces for marrow. Though spongy bone is found in the ends of long bones, it occurs in the middle of all other types of bones. Since the spongy tissue is weaker than the harder bone of the shaft, a protective layer of compact bone always surrounds the spongy bone on the ends of long bones.

There is a trade-off in having very hard bones. The harder a substance is, the more likely it is to fracture under a sudden, strong blow. This is due to the inability of compact bone to absorb the energy of the blow. The flat bones of the skull are made in a sandwich-like fashion. They have inner and outer layers of compact bone, with a layer of spongy bone, called **diploe**, in the middle. The spaces in the spongy bone allow the energy of a blow to the head to be absorbed. While the outer layer of compact bone may fracture, only a tremendously hard blow would have sufficient energy to work through the spaces of the spongy bone and fracture the inner compact bone layer. The diploe layer thus acts as a shock absorber.

SECTIONS OF A LONG BONE

As noted previously, the shaft of a long bone is also called the diaphysis, and on either end of the diaphysis, the bone swells outward to form an epiphysis. Enlargement of the epiphysis adds strength to the joint, where much of the stress is placed on the skeletal system. The epiphysis is also the place where ligaments and tendons attach. The diaphysis provides leverage, increasing the power or the range of the muscle movements (Figure 6.1).

You may recall that the term *articulate* describes the point at which bones meet. At this point, the surface that forms the joint is covered by a type of hyaline cartilage called **articular cartilage**. This cartilage works with the lubricating fluids of the joint to reduce friction and to increase the efficiency of movement at the joint.

Bones are covered with a tough, outer sheath called the periosteum. The periosteum consists of two layers. The outer layer is a tough, fibrous sheet of **collagen**. The inner layer, called the osteogenic layer (*osteo* means "bone" and *genic* means "to create"), contains the cells that allow bone to grow, dissolve, and repair.

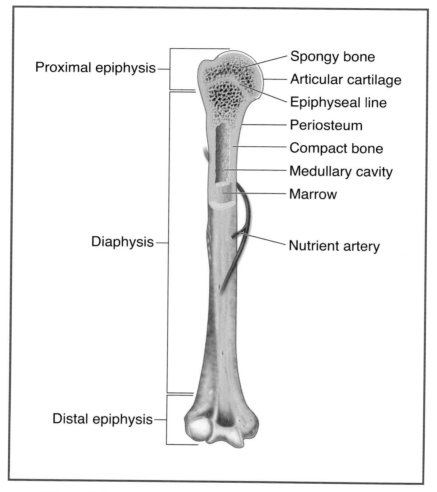

Proximal epiphysis

Spongy bone

Articular cartilage

Epiphyseal line

Periosteum

Compact bone

Medullary cavity

Marrow

Diaphysis

Nutrient artery

Distal epiphysis

Figure 6.1 As mentioned earlier, bones are not completely solid structures. Instead, they are composed of hollow and solid structures. This allows them to be strong enough to support the body, but also provides routes for vital nutrients and blood to flow through them.

There is also a membrane inside the long bone called the endosteum. The endosteum is made of a thin layer of connective tissue that contains **osteogenic cells**, or bone-growing cells. As a result, bones grow from both the outside and the inside.

The composition of our long bones changes as we age. In young children, the epiphyses are separated from the diaphysis by plates of hyaline cartilage where bones grow in length. When we reach our maximum height, this plate is converted to compact bone and only a thin line remains as evidence that the cartilage plate once existed.

BONE CELLS

Bone tissue is a connective tissue and, like all connective tissues, it contains a mixture of ground substance, fibers, and living cells. There are four basic types of bone cells: osteogenic cells, osteoblasts, osteocytes, and osteoclasts (Figure 6.2). Osteogenic cells are found in the endosteum, the inner layer of the periosteum, and in the central canals of bone. Osteogenic cells continually undergo mitosis, and some of the cells produced develop into a second type of bone cells called osteoblasts. Osteogenic cells are **stem cells**; that is, they are continually giving rise to more osteogenic cells and osteoblasts.

Osteoblasts are the cells that form bone. An osteoblast cannot undergo mitosis and, therefore, cannot reproduce. Osteoblasts are only produced by conversion of osteogenic cells through a process known as **differentiation**. They produce the organic material found in bone and help to mineralize the bone with calcium phosphate. Osteoblasts are found in rows in the inner layer of the periosteum and in the endosteum. These cells are critical for bone growth and bone repair. Not surprisingly, stress to bone or a fracture to a bone will stimulate the reproduction of osteogenic cells, thus giving rise to a larger number of osteoblasts for building and repair.

Osteocytes are osteoblasts that become trapped in the matrix the osteoblasts produce. They are found in little spaces called **lacunae**. The primary role of these cells is to detect changes in bone and to communicate those changes to neighboring osteoblasts and osteoclasts. They also assist with transferring nutrients and removing waste products from bone.

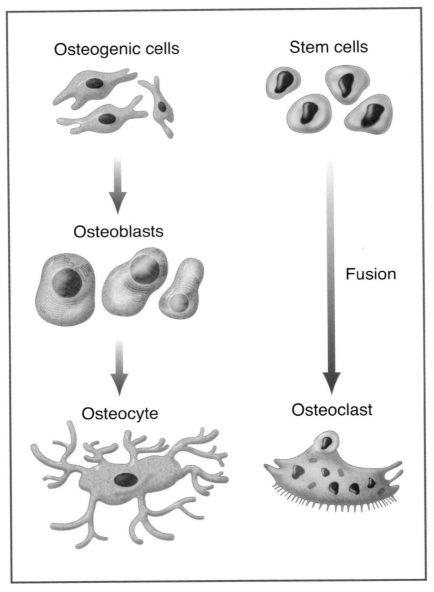

Figure 6.2 The formation of osetoblasts, osteocytes, and osteoclasts is illustrated in this diagram. Osteogenic cells give rise to
osteoblasts, which further develop into osteocytes. Stem cells fuse
to form osteoclasts.

Signals to osteoblasts trigger these cells to build bone, while signals to osteoclasts trigger the removal of bone.

Osteoclasts are cells that remove bone. Osteoclasts are made from bone marrow stem cells that fuse together. Thus, they are relatively large cells and have more than one nucleus (a trait common to fused cells).

BONE MATRIX

Bone is made of a matrix of materials, one-third of which is organic (carbon-based) and two-thirds of which are inorganic. The organic material in bone is primarily a mix of collagen and **glycoproteins**, proteins modified with carbohydrates. Calcium phosphate in a form called **hydroxyapatite** makes up about 85% of the inorganic portion of the matrix, with the remaining 15% composed of a variety of other minerals and ions. Since bone is made of both organic material and inorganic material, it is said to be a **composite material**. The organic portion of bone gives it flexibility, while the inorganic portion provides strength. Both components are essential to function. If bone is depleted of inorganic components, particularly calcium phosphate, it becomes too flexible, almost like rubber. A nutritional disease called **rickets** results from a deficiency of vitamin D, a vitamin that promotes calcium absorption by the body. When the body is deficient in vitamin D, it cannot regulate calcium and phosphate levels. If the blood level of these minerals becomes too low, calcium and phosphate are released from the bones into the bloodstream in order to elevate the blood levels. Rickets causes a progressive softening and weakening of the bones that eventually can lead to deformities. These are most noticeable in the long bones of the legs, since they bear the weight of the body (Figure 6.3). A condition called **brittle bone disease** results when the protein component of bone, primarily collagen, is too low in concentration. This results in a loose and very weak bone structure that does not contain

HOW BONES GET WEAK: AN EXPERIMENT

Rickets is a disease caused by a lack of vitamin D, which is needed for the body to absorb calcium. When a person has rickets, the bones weaken and become deformed. You can demonstrate the effects of rickets by doing this simple experiment. The next time you have chicken for dinner, set aside the bones from the two drumsticks. Clean the bones as best you can to remove meat and other loose tissue.

Before you proceed, hold each of the drumstick bones (these are the femurs of the chicken) with one thumb and forefinger on each end. Apply a little pressure to make sure that the bones are still solid.

Next, place each of the bones into a separate tall drinking glass or jar. Pour milk into one of the jars until the bone is completely covered. This will be your control. Then, pour vinegar into the container with the other bone until the bone is fully immersed. Vinegar is an acid that will extract, or leach, calcium phosphate from the bone. Cover each container with plastic wrap and label them properly (this is not only good scientific practice, but will also keep someone from drinking your experiment!). Place the containers in the refrigerator for five days.

After the five days, remove both containers from the refrigerator and pour out the liquid. Wash both of the bones with running water until they are clean. First, check the rigidity of the control bone, the one that was soaked in milk. How does it compare to the bones with which you started? Next, test the rigidity of the vinegar-soaked bone. You should notice a substantial difference between the two bones. The one soaked in vinegar should be much softer and weaker, because the vinegar has caused the bone to demineralize. Finally, stand each bone on end and press down from the top with your thumb. Which bone bends under pressure? What happens to the vinegar-soaked bone is similar to what happens to the leg bones of a person who has rickets!

Figure 6.3 The boy pictured here has rickets, a disease resulting from a lack of calcium and/or vitamin D in the diet. Rickets weakens the long bones, so that they are not strong enough to support the body; they bend outwardly under the body's weight.

enough calcium and breaks easily. In persons with this disease, bones can shatter under weight or stress.

BONE MARROW

The soft tissue found inside the cavities of bones is called bone marrow. There are three types of bone marrow: red marrow, yellow marrow, and gelatinous marrow. The ratio of these three types of marrow changes with age and reflects the demands of the body at any given age.

Red bone marrow is also called **hemopoietic tissue**, or the tissue that gives rise to blood cells. Just about all of the bone marrow in young children is of this class and reflects the need to produce many blood cells while the body is growing and developing.

Yellow bone marrow contains much more fatty tissue than red bone marrow does. By the time a person reaches middle age, the majority of red bone marrow has become yellow bone marrow. While yellow bone marrow cannot produce blood cells, in an emergency, such as in cases of anemia, a condition in which the body does not have enough red blood cells, it can be converted back to red bone marrow. In adults, most of the red bone marrow that remains is found in the ribs, the vertebrae, the sternum, parts of the pelvic girdle, the heads of the femur where it joins the hip, and the humerus where it joins the shoulder. The primary function for yellow bone marrow may be fat storage.

By the time we reach old age, most of our yellow bone marrow has become **gelatinous bone marrow**. Gelatinous marrow looks like reddish-brown jelly. We do not know of a function for this form of marrow.

OSSIFICATION

The process of first producing bone is known as ossification (Figure 6.4). Generally, this process involves the conversion of hyaline cartilage to solid bone. This process begins with invasion of the cartilage by blood vessels in what will be the diaphysis of the bone, then later, in the epiphyses. Between the two, on each end of the bone, a new region, the metaphysis, is formed. Most bone growth occurs in the metaphysis. Each metaphysis contains five clearly distinct zones, all of which are critical to bone formation:

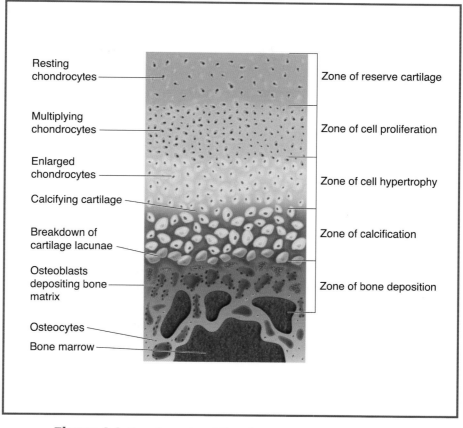

Figure 6.4 Bone is produced though a process called ossification. Most of this growth occurs in the metaphysis, which contains five different growth zones. The active zones during bone growth and deposition are illustrated here.

1. The **zone of reserve cartilage** is the region farthest away from the center of the developing bone. It contains the cartilage that has yet to begin the conversion process.

2. The **zone of cell proliferation** is the site toward the center of the bone that is nearest the zone of reserve cartilage. In this zone, cells called chondrocytes, or cartilage cells, multiply.

3. The **zone of cell hypertrophy** is the site where chondrocytes stop dividing. Zones two and three are continually pushing toward the end of the bone, reducing the size of the zone of reserve cartilage.

4. The **zone of calcification** is where minerals begin to deposit between columns of lacunae and is where the calcification of cartilage occurs.

5. The **zone of bone deposition** is where the lacunae begin to break down and the chondrocytes begin to die off. As the lacunae break down, hollow channels are formed that fill with marrow and blood vessels. Osteoclasts dissolve the calcified cartilage, while osteoblasts begin to deposit concentric rings of matrix. As the concentric layers are added, the channels get smaller and smaller. Eventually, the osteoblasts become trapped in the tiny channels and convert to osteocytes, which stops the process of making additional layers of matrix.

REPAIR OF BONE BREAKS AND FRACTURES

Our bones do not form and remain unchanged. Stresses will continually cause us to reform, reshape, and grow the amount of bone. For instance, the stress of bearing weight as we begin to walk upright signals the body to increase bone production. Lifting weights and other forms of weight-bearing exercise will increase the bone mass in our bodies.

In some cases, our bones may fracture or break. The body has developed a mechanism to repair damaged bone. This process occurs in four stages (Figure 6.5).

The first stage is the formation of a **hematoma**, the site of internal bleeding. When a bone fractures, blood vessels in the area are also damaged and blood leaks into the surrounding tissues. The hematoma develops into a blood clot.

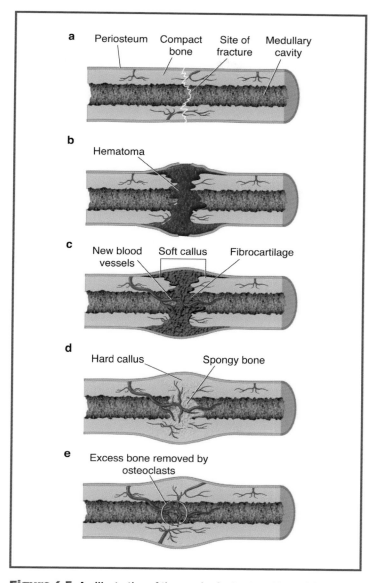

Figure 6.5 An illustration of the repair of a fractured bone (a) is shown in this diagram. Blood infiltrates the damaged site, forming a hematoma (b), a soft callus of fibrocartilage forms around the hematoma to provide support (c), osteoblasts produce a hard callus to strengthen the soft callus (d), and finally, osteoclasts remove excess bone and callus (e).

The second stage in bone repair is the formation of **granulation tissue**. As blood vessels infiltrate the hematoma, a soft, fibrous tissue is laid down. The site also attracts macrophages (a type of immune cell), osteoclasts, and osteogenic cells. The osteogenic cells reproduce rapidly within two days of initial injury.

The third stage in the process is **callus formation**. Several cells are important in this process. Fibroblasts deposit collagen in the granulation tissue. Some of the osteogenic cells differentiate into **chondroblasts**, or collagen-producing cells, which form areas of soft callus tissue made of fibrocartilage. Other osteogenic cells differentiate into osteoblasts, which produce the **hard callus**, a bony collar. The hard callus adheres to the dead bone around the site of injury and acts as a temporary splint to hold the broken ends together. This is why doctors attempt to "set" a broken bone as soon as possible after an injury. Setting the bone positions the broken ends in the most natural position for healing. If the bone is not properly set, the repaired bone may be deformed. It takes about 4 to 6 weeks for the hard callus to form. Because the bone shouldn't move during that time, a broken bone is usually put into a cast or splint to keep the "set" bone in place.

The final step in the process is called **remodeling**. During remodeling, which requires 3 to 4 months, hard callus remains at the site. Osteoclasts dissolve small fragments of bone, and osteoblasts bridge the gap between the broken bone with spongy bone. Eventually, this spongy bone is remodeled to compact bone. Within 4 to 6 months, repair is complete. Generally, the older we are, or the poorer our general health, the greater the length of time for the healing of bone.

CONNECTIONS

In this chapter, we have explored the nature of bone tissues, bone cells, and the process for growing and repairing bones. Using long bones as a model, we have explored the structure of bone and have learned about the processes for making new bone and repairing broken or fractured bones. We have seen that bones consist of both hard and soft tissues and that both types play a vital role in the function of the skeletal system. We have also seen that different bones contain different types of marrow and that the distribution of the different types of marrow changes with age. Finally, we have explored the role of various cells in bone growth, both during natural growth and development of the body, and in the repair of broken or otherwise damaged bones.

7

Muscles, Muscle Cells, and Muscle Tissues

MUSCLES, MUSCLE CELLS, AND MUSCLE TISSUES
Types of Muscles

The study of muscles is called **myology**. Muscles can be divided into three primary types: skeletal muscle, smooth muscle, and cardiac muscle. Skeletal muscles are those that are usually attached to bone and are responsible for movement and stability. The skeletal muscles will be the focus of Chapter 8. The smooth muscles are associated with other organ systems of the body, such at the circulatory, digestive system, and urinary systems, where conscious muscle control is neither required nor desirable. Finally, cardiac muscles are associated with the heart. These muscles have physical characteristics similar to skeletal muscle, but they are not under the body's conscious control (Figure 7.1).

CHARACTERISTICS OF MUSCLES AND MUSCLE TISSUES

Even though there are three distinct types of muscles, all muscles have certain characteristics in common. These five traits of muscle cells help to distinguish them from other cell types in the body.

All muscles demonstrate **responsiveness**. Although responsiveness is a property of all living cells, it is particularly notable among muscle cells. As muscle cells are stimulated by chemical signals,

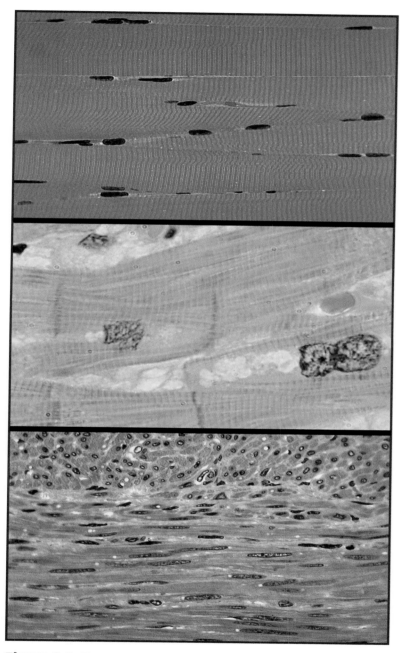

Figure 7.1 Shown here are micrographs of skeletal muscle (top), cardiac muscle (middle), and smooth muscle (bottom). Each type of muscle has a different function.

stretching, electrical charge, or other stimuli, the cells respond through an electrical change across their plasma membrane of the cell.

A second common trait of all muscles is **conductivity**. When one muscle cell is stimulated, the electrical charge it generates across its membrane stimulates the muscles on either side of it, causing the charge to travel along the muscle cells and to communicate the response throughout the muscle tissue.

The third common trait of muscle cells is **contractability**. When stimulated, muscle cells have the ability to shorten or contract. When this is done in concert with a number of other muscle cells in the same region, the muscles "pull" on bone or other tissues, resulting in movement.

The fourth common characteristic of muscle cells is **extendibility**. This is the opposite of contractability, in that the muscles must extend between contractions. Some muscle cells are three times longer when extended than when contracted.

Finally, all muscle cells must have **elasticity**. When muscles are relaxed, they can stretch. When the tension causing the muscles to stretch is released, an elastic cell returns to its original length and shape. Very few other cells in the human body have the ability to survive stretching.

SKELETAL MUSCLE
Characteristics

Skeletal muscles have three defining characteristics. First, they are **voluntary**, meaning that we can consciously control their movements and functions. The second important characteristic is that skeletal muscles are **striated**, or have visible stripes or lines when viewed under the microscope. The third trait of skeletal muscles is that they are usually attached to one or more bones. A typical skeletal muscle is about 3.0 cm long and about 100 μm in diameter. Since these cells are so long, they are often referred to as muscle fibers. Involuntary muscles are not attached to bone and are not under conscious control; thus, they can be distinguished from skeletal muscle. They also lack the striations common to both voluntary and cardiac muscle.

Skeletal muscles consist of both skeletal muscle tissue and fibrous connective tissue. Each individual muscle fiber is surrounded by the **endomysium**. Bundles of muscle fibers that work together to conduct functions are called **fascicles**. Fascicles are surrounded by a connective tissue sheath known as the **perimysium**, and entire muscles are surrounded by connective tissue known as **epimysium**. These connective tissues join the collagen fibers of the tendons, and tendons join the collagen of bone matrix, physically linking muscles to bones. As a result, if a muscle contacts, it pulls on the connective tissue, which causes a bone to move.

Collagen, a soft tissue of the body, shares some physical properties with muscle, while differing in other properties. Collagen shares with muscle the characteristics of extensibility and elasticity, but lacks responsiveness, conductivity, and contractability. As a result, collagen can be stretched and restored to its original form, but it cannot do so actively. Muscle must drive changes in collagen configuration.

Composition

During embryonic development, muscle cells develop from special stem cells called **myoblasts**. The myoblasts fuse together to form each muscle fiber, and the nuclei fuse to form the long, cylinder-shaped nuclei positioned just below the plasma membrane that are characteristic of skeletal muscle cells. A few of the myoblasts become **satellite cells** that remain associated with muscles and multiply and produce new muscle fibers if muscles become injured and need repair. The plasma membrane of a skeletal muscle cell is called the **sarcolemma**, while the cytoplasm is known as **sarcoplasm**. The sarcoplasm is filled with protein bundles called **myofibrils** that are about 1 μm long. The sarcoplasm also contains large amounts of **glycerol**, a three-carbon alcohol that serves as a source of energy for muscle cells, and a protein called **myoglobin**, which stores oxygen for metabolism. Finally, the smooth endoplasmic reticulum of muscle cells is called the **sarcoplasmic reticulum**. This forms a network around each myofibril with sacs called **terminal cisternae**, which are storage sites for calcium ions. The controlled release of these calcium ions through gates controls muscle contraction.

Myofilaments

The structural units of myofibrils are called **myofilaments** (Figure 7.2). There are three distinct types of myofilaments. Thick filaments are about 15 nm in diameter and are made up of several hundred myosin molecules. Each myosin molecule is made of two polypeptides and is shaped like a golf club. A typical myofilament contains between 200 and 500 of these units.

Thin filaments are about 7 nm in diameter and are made up of two molecules of a protein called **fibrous** or **F actin**. Each F actin is like a string of beads. These beads are called **globular** or **G actin**. Each G actin can bind to the head of a myosin molecule. The thin filaments each have 40 to 60 molecules of a different protein, **tropomyosin**. When a muscle is relaxed, each tropomyosin molecule blocks several G actin molecules, preventing binding with the heads of the myosin molecules.

Elastic filaments are about 1 nm in diameter and consist of a protein called titin. Elastic filaments run down the core of thick filaments. The elastic filaments extend from the thick filaments on the ends and serve to anchor the filaments to Z disks. Z disks and elastin fibers are important in keeping the muscle fibers together, preventing them from overstretching, and play a role in allowing the cells to return to their resting length after being stretched.

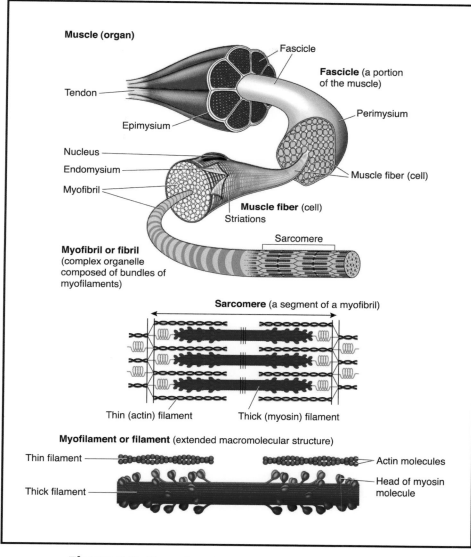

Figure 7.2 Shown here is an illustration of the organization of skeletal muscles. Muscles contain myofibrils, which are fibers composed of bundles of myofilaments. There are two types of myofilaments: thick and thin. Thick myofilaments are made up of myosin molecules. Thin myofilaments are composed of F actin, G actin, and tropomyosin.

Myosin and actin are the contractile proteins of muscles, while tropomyosin and **troponin** (a calcium-binding protein) are the regulatory proteins.

Striations

All skeletal muscles have alternating dark and light bands (Figure 7.3). These bands form the characteristic striated pattern of skeletal muscles and cardiac muscle. The dark band, which corresponds to the length of the thick filaments, is called the A band, while the light band, where there are only thin filaments, are called I bands. Within the A band is an especially dark region. This is the point where the thick and thin filaments overlap. In the middle of the A band is a lighter region, the H band, which contains thick filaments but not thin ones.

I bands also have a dark line that runs down the center. This dark line is made of Z disks. Recall that the Z disks serve as the anchoring site for the thick filaments and the elastic filaments. Between each Z disk is a region of the myofibril called a **sarcomere** (Figure 7.4), the contractile unit for the muscle. When a muscle contracts, it does so because the sarcomeres shorten, drawing the Z disks closer together. The Z disk is also connected to the sarcolemma. When the Z disks are pulled together, the tension on the sarcolemma makes the entire muscle cell shorten.

WHAT CAUSES MUSCLES TO MOVE?

We have examined the basic composition of muscle cells and have noted that contraction of muscles, at least skeletal muscles, leads to movement, but what causes muscle fibers to contract? In this next section, we will explore some of the more common signals that cause muscles to work.

Nerve Stimulation

When a skeletal muscle remains unstimulated, it will not contract. The most common stimulus for a muscle is stimulation

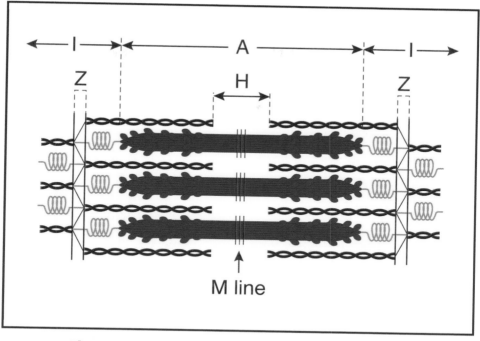

Figure 7.3 The organization of actin and myosin is responsible for the visible striations (alternating light and dark bands) in skeletal muscle. Each band is assigned a letter designation, depending on the types of cells and filaments it contains.

by the nervous system. A signal along a nerve channel is basically electrical, while chemical signals at the neuron junctions keep the electrical signal moving. Muscles get their signals from **somatic motor neurons**, found in the brainstem and spinal cord. These impulses or signals are carried to the muscles through the axons, or tails, of these somatic motor neurons. The ends of these axons branch many times. Each branch leads to a different muscle fiber, and each muscle fiber can only be stimulated, or inner-vated, by one neuron (Figure 7.5). When a motor neuron sends out a signal to a muscle, an electrical charge travels down the branches and each signal stimulates about 200

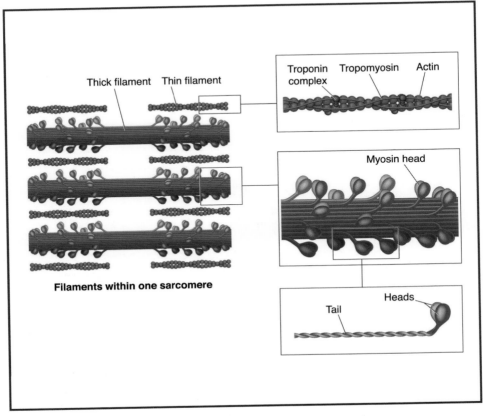

Figure 7.4 The composition of a single sarcomere, the contractile unit for the muscle, is illustrated here. Muscles contract and relax as the myosin heads move back and forth. This is discussed in more detail on page 101.

muscle fibers, which are coordinated to contract together. All muscle fibers stimulated to contract from a single neuron form a **motor unit**. This allows one nerve to coordinate an entire muscle.

What happens if the signal cannot get from the nerve to the muscle group? This might happen for a number of reasons. If the neuron is damaged, it may not transmit its charge to the muscles. Some poisons prevent the neuron from

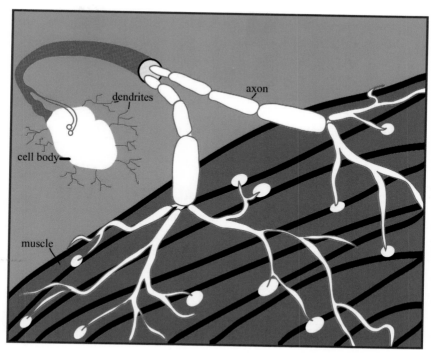

Figure 7.5 A single nerve is able to stimulate multiple muscle fibers. Nerve cells, or neurons, are composed of dendrites that receive information, a cell body, and an axon, which passes on the information. Axons can branch many times and connect to many muscle fibers; thus, one signal can stimulate more than one muscle.

communicating with the motor group. In any case, if the signal doesn't get through, the muscle will not contract. Remember that the signal is electrical. You can make a simple model for the nerve/muscle interaction using a battery, a wire, a switch, and a light bulb. When the circuit is complete, electrons flow from the battery through the circuit, including the light bulb, and back to the battery again. If you break the wire along the path, the current cannot flow and the light bulb will not light. If you open the switch (break the contacts so current does not flow), the same thing happens. Damage to the nerve or blocking the communication between the nerve and the muscle has the same effect as breaking the circuit in the electrical model, in that the muscle will not contract if the circuit is broken.

Electrical Stimulation

Muscles can be stimulated to contract by using a mild electrical current. Since nerves communicate with muscles electrically,

HOPE FOR HEALING SPINAL CORD INJURIES

Although spinal cord injuries are technically damage to the nervous system and not the muscular system, the constant need for communication between the two systems means that injuries affect them both. Spinal cord injuries are among the most severe. Since the spinal cord is made of nervous tissue, which is not able to regenerate, an injury to the spinal cord is currently permanent. Though we may not be able to reverse spinal injury, modern technology and ongoing medical research offers new hope for overriding the damage to the spinal column.

As our understanding of muscle physiology, electrical engineering, and computer miniaturization develops, there may be hope for those who are paraplegic (paralyzed from the waist down) or quadriplegic (paralyzed from the neck down). New techniques may someday allow us to develop artificial "nerve channels" using wires and electrodes. With the help of computer chips complex enough to control thousands of functions at once, it may be possible to program a computer to assist people who have spinal cord injuries in stimulating the muscles of the legs, arms, or chest. Such a program would essentially bypass the spinal column and communicate directly with the muscles. More advances have been made in medicine in the past ten years than in all of history before that. Robotics, nanotechnology, and human engineering may not only open new doors, but they may actually let some people who are currently in wheelchairs walk through them!

artificially applying an electrical charge across a muscle will also cause it to contract. Doctors sometimes do this to test muscle responsiveness.

Calcium Ion Stimulation

You can short-circuit a nerve channel by the addition of certain chemicals. Calcium ions, (Ca^{++}) for example, are stored in neurons. When electrical signals reach the gate that keeps calcium ions in, the gate opens and the calcium ions flood into the synaptic knob of a neuron. The increased calcium ion then causes the neuron to release a chemical, acetylcholine, into the synaptic junction, which stimulates the receptors on the next neuron in that nerve channel. The stimulation causes an electrical signal to travel down the neuron to the end, where the synaptic signal is repeated. This is how the electrical current travels from one neuron to the next or from a neuron to a muscle fiber. Treating tissues with calcium ions will cause artificial contraction of muscles.

THE MECHANICS OF MUSCLES CONTRACTION

We have already noted that contraction of muscles involves shortening of the sarcomere, but how does this happen? Remember that the contraction of a muscle is movement, and movement requires expenditure of energy. The main energy currency for a living cell is **ATP (adenosine triphosphate)**. ATP holds the last two of its three phosphate groups onto the molecule using high-energy bonds, so if those bonds undergo hydrolysis, the splitting of a compound into fragments by the addition of water, energy is released. It may not surprise you to learn that contraction of muscles involves the hydrolysis of ATP.

The "head end" of a myosin molecule has a binding site for ATP (Figure 7.6). In order for myosin to begin the contraction of the sarcomere, ATP must be attached. This binding

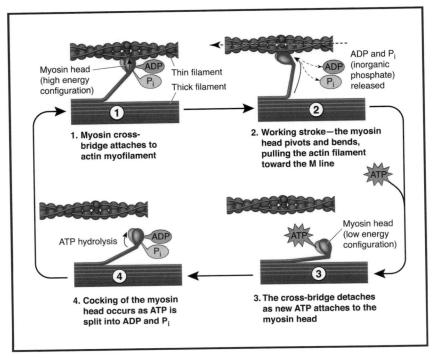

Myosin head (high energy configuration)

Thin filament
Thick filament

1. Myosin cross-bridge attaches to actin myofilament

ADP and P$_i$ (inorganic phosphate) released

2. Working stroke—the myosin head pivots and bends, pulling the actin filament toward the M line

Myosin head (low energy configuration)

3. The cross-bridge detaches as new ATP attaches to the myosin head

ATP hydrolysis

4. Cocking of the myosin head occurs as ATP is split into ADP and P$_i$

Figure 7.6 The process of muscle movement is driven by the energy released from adenosine triphosphate (ATP). The energy released when ATP is hydrolyzed to ADP and phosphate positions the myosin head in the high-energy position. As ADP and phosphate are released, the myosin undergoes the motion stroke. ATP binds to myosin, returning it to the low-energy state, and as the ATP is broken down into ADP, phosphate, and energy, myosin is energized to make contact with the thin filament.

site has an enzyme near it called myosin ATPase. As the ATPase hydrolyzes ATP, two by-products—ADP (adenosine diphosphate) and phosphate ion—are generated but remain attached to the myosin head. The energy released from ATP is absorbed by the myosin, which moves into an "activated" or "energized" form. As the ADP and phosphate are released, the myosin goes to the "relaxed" form and, as it does, it pulls along the actin chain associated with the myosin head. The head of the myosin remains attached to the actin until a new ATP binds myosin, which causes it to release. As this process is repeated over and over, the myosin crawls along the actin chain toward the middle of the sarcomere. Since myosin heads extend toward the center of the sarcomere from each side, the two ends pull toward the middle, shortening the chain.

A mechanical model for this myosin/actin association is a bicycle chain and its cog. Think of the teeth on the cog as myosin and the chain as actin. When you put energy into the cog, a tooth moves forward, drawing the chain along with it. Add more energy and the chain pulls farther. Now imagine two bicycles connected to the same chain. If you put energy into both by pressing on the pedals, the bicycles will be drawn toward each other, shortening the distance between them.

OTHER TYPES OF MUSCLE
Cardiac Muscle

Cardiac muscle is found only in the heart. Like skeletal muscle, it is striated, but for a different reason. Cardiac muscle striations are due to intercalated discs, which form the thick, dark lines in the muscles. These are electrical gap junctions that allow each cell (called a **myocyte**) to stimulate its neighboring myocytes electrically. This means that cardiac muscle can be stimulated to contract without a signal from a nerve. The heart contains a specialized structure called a **pacemaker** that produces and releases electrical charges at a fixed rhythm. This electrical charge travels over the heart in a wave, causing the various chambers to contract. This squeezes blood through the heart and drives blood circulation.

Another important difference between cardiac muscle and skeletal muscle is the number and size of the mitochondria, the cell's factories for making ATP. The large mitochondria of cardiac muscle make up about 25% of the volume of the muscle, whereas the smaller mitochondria of skeletal muscles make up only about 2% of the volume. The cardiac muscle mitochondria use a wider range of energy sources than skeletal muscle mitochondria, but are much more sensitive to the depletion of oxygen. Most heart attacks occur when the oxygen supply to the cardiac muscle is interrupted. The energy level in the cardiac tissue drops and the heart ceases to function normally.

Smooth Muscle

Smooth muscle can be divided into two types: multiunit smooth muscle and single-unit smooth muscle. Multiunit smooth muscle is located in the walls of the large arteries, in the pulmonary air passages, in the iris of the eye, and in the piloerector muscles of hair follicles. A separate terminal branch from the nerve (in this case, an autonomic nerve) is attached to each cell, as in the skeletal muscles discussed earlier, so that a signal travels to a cluster of muscle cells or a motor unit.

A more common mechanism for control of smooth muscle is the single-unit smooth muscle, which is found in the digestive, urinary, respiratory, and reproductive tracts, as well as in most blood vessels. With single-unit smooth muscle, two nerve patterns exist. The inner track innervates myocytes that form a circle around the organ, while the outer track is longitudinal, with myocytes running lengthwise along the organ. These myocytes are considered single-unit because they don't each have a nerve branch. Instead, they have gap junctions (like cardiac cells) that allow myocytes to stimulate each other. This means that a large number of cells contract together, almost as if they constituted a single muscle cell.

Although smooth muscle, like cardiac muscle, is controlled by the autonomic nervous system, and we therefore have no control over the process, there are stimuli that affect the contraction of smooth muscle. For instance, some of the smooth muscles of the respiratory system and circulatory system are stimulated by concentrations of the gases oxygen and carbon dioxide. Acid levels affect contraction of smooth muscle in the digestive system. Still other smooth muscle may be stimulated by hormones or stretch responses (in the stomach and bladder).

CONNECTIONS

Three basic types of muscles exist in the human body. Skeletal muscle is generally under our conscious control and has a striated appearance. Cardiac muscle is also striated, but is under the control of the autonomic nervous system and, thus, does not require our conscious intervention to function. Smooth muscle lacks striations and is also under the control of the autonomic system.

All muscles contain myosin and actin and use ATP as the source of energy for movement. All muscles, whether skeletal, cardiac, or smooth, share the same basic mechanism of contraction.

8

Skeletal Muscles:
Form and Function

GENERAL MUSCLE ANATOMY

Muscles come in a variety of types and perform a variety of functions. As a general rule, most muscles attach to bones or other structures at two sites. One attachment site is relatively immobile, while the other attachment site is much more mobile. The relatively immobile, or stationary, end of a muscle is called the **origin**, while the more mobile end of the same muscle is known as the **insertion**. Many muscles tend to be wider in the middle and taper toward the origin and the insertion. This thickened middle area is called the **belly**.

The more than 600 muscles of the human body can be divided into 5 groups based on the arrangement of the bundles of muscle fiber. The groups are fusiform, parallel, convergent, pennate, and circular muscles (Figure 8.1).

Fusiform Muscles

Fusiform muscles are spindle-shaped: tapered on the ends and thicker in the middle. Since this design concentrates the strength of a large bundle of muscles at relatively small origins and insertions, fusiform muscles tend to be strong. The biceps muscles of the upper arm are examples of fusiform muscles.

Parallel Muscles

As the name suggests, **parallel muscles** have bundles of fascicles that are essentially equally wide at the origin, the insertion, and the belly.

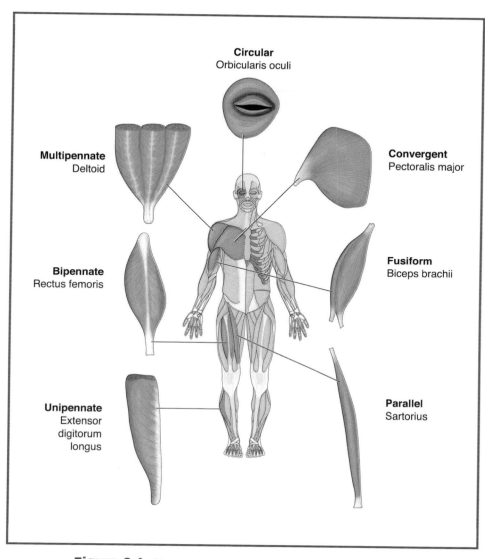

Figure 8.1 Muscles can be divided into five groups: fusiform, parallel, convergent, pennate, and circular. Their unique shapes and the ways they connect to the bones are responsible for the wide range of movement of which the human body is capable.

These muscles form belt-like structures. In the relaxed state, parallel muscles can be quite long. Although they can shorten more than fusiform muscles, relative to their relaxed length, they tend to be less strong than fusiform muscles. The muscles of your abdomen are examples of parallel muscles.

Convergent Muscles

Convergent muscles are shaped like a fan. They are wide at the origin and narrow at the insertion. The strength of a large number of fascicles concentrated at the insertion makes convergent muscles quite strong. The pectoralis major muscles of the chest are examples of convergent muscles.

Pennate Muscles

Pennate muscles are shaped like feathers. The bundles of muscle fibers insert onto a tendon that runs the length of the muscle. Pinnate muscles can be further divided into three groups. **Unipennate muscles** are those in which the fascicles all attach from the same side. The palmar interosseous muscles in your palm are examples of unipennate muscles. **Bipennate muscles** have bundles of muscle fibers that attach to the tendon from two sides with the tendon in the middle. The rectus femoris muscle of the thigh is an example of a bipennate muscle. **Multipennate muscles** are shaped like several feathers that have all their quills joining at a single point. An example of a multipennate muscle is the deltoid muscle of your shoulder.

Circular (Sphincter) Muscles

Circular or **sphincter muscles** form rings around various openings in the body. The muscles around the lips and the eyelids are examples of circular muscles.

WORKING IN GROUPS

Body movements usually do not involve the action of a single muscle alone. More often, groups of muscles coordinate their actions to provide a smooth, even movement. As a general rule, the muscles that work together as a functional group fall into four major categories.

Agonist Muscles

The muscle that exerts the majority of force in a movement is known as the **agonist** or **primary mover**. When you bend your elbow, the biceps muscle is the agonist.

Synergist Muscles

Most agonist muscles have other muscles that aid the motion made by the agonist. Muscles that work additively to the agonist are known as **synergist** muscles. In addition to adding strength to the agonist, the synergist helps to stabilize the movement or to restrict the range of movement of the agonist.

Antagonist Muscles

It is important to remember that muscles can only pull, they cannot push. In order to reverse the action of the agonist, a complementary muscle or group of muscles must work in the opposite direction. An **antagonist** muscle works in opposition to the agonist. In addition to returning the body part to the original position, antagonists fine-tune the control of the agonist. They moderate the speed and the range of the agonist, which helps to protect the body from damage to muscles or joints. The triceps muscle along the back of the upper arm is the antagonist to the biceps. It is important to remember that while the biceps is the agonist and the triceps is the antagonist when we bend the elbow, their roles reverse when we straighten our elbow; that is, the triceps becomes the agonist and the biceps the antagonist.

Fixator Muscles

Fixator muscles are those that prevent the bone from moving in an unwanted direction. For instance, when you want to bend your elbow, the biceps muscle does the majority of the work. The biceps connects to the shoulder blade or scapula at its origin and on the radius at its insertion. Fixator muscles attached to the scapula prevent it from moving when the biceps contracts, ensuring that the energy of the biceps concentrates on moving the radius, not the shoulder blade.

MUSCLES OF THE FACE

Generally, the muscles of the face are small, short muscles that allow tremendous precision and control, but not a significant amount of strength. The exceptions to this rule are the muscles of the jaw, which are extremely strong. Facial muscles control facial expression, chewing and swallowing, and, in part, speech.

Facial Expression

The muscles that control facial expression are much more highly developed in primates than they are in other animals. The mouth is the most sensitive structure to facial expression. It is surrounded by a circular muscle responsible for closing the lips. A similar circular muscle surrounds each eye and allows the eyelids to close. The other muscles of the mouth radiate outward from the mouth like spokes on a wheel. The origins of the muscles tend to be away from the mouth, while the inserts are attached to the skin rather than to bone. This combination of specialized structures and anchoring to soft tissue allows for countless and subtle facial expressions, from smiles to frowns to pouts.

Chewing and Swallowing

The tongue is one of the most flexible and strong structures in the human body. It consists of sets of muscles originating and terminating within the tongue (intrinsic muscles), and muscles that connect the tongue to other parts of the neck and head (external muscles). The tongue plays a vital role in digestion, as it, along with the muscles of the cheeks, moves food around the mouth. The tongue also delivers the chewed food to the back of the throat, where it enters the esophagus.

The action of chewing is controlled by four paired muscles of **mastication**, or the act of biting and grinding food in your mouth. Two of these that attach at or near the mandible are among the strongest muscles of the body. They allow us to bite and to chew our food. The other two pairs of chewing muscles are important in moving the mandible from side to side, allowing for the grinding of food by the molars. Also important for chewing and swallowing are eight pairs of **hyoid** muscles, associated with the hyoid bone, a U-shaped bone at the base of the tongue that supports the muscles.

As food enters the pharynx, the back of the throat, three pairs of muscles known as the pharyngeal constrictors contract and force the food down the esophagus.

The larynx also has intrinsic muscles that control the vocal cords, and thus, are essential for speech. These muscles also control the opening of the larynx itself, preventing food or fluids from entering the trachea.

Muscles That Control the Head

The bones of the skull are very heavy. Consequently, the muscles that control and support the head must be very strong. There are two major groups of muscles for the head: flexors and extensors. All muscles in these groups have their inserts on cranial bones and their origins on the vertebrae, the thoracic cage, or the pectoral girdle. The **flexors** draw the head downward toward the body. These muscles are generally located along the sides of the neck. The **extensors** extend the neck and rotate the head. The extensor muscles are located at the back of the neck.

MUSCLES OF THE TRUNK

The muscles of the trunk can be divided into three groups: the muscles of the abdomen, the muscles of the back, and the muscles of respiration.

Muscles of the Abdomen

The abdominal muscles consist of four pairs of muscles that form strong sheets. These muscles serve a number of functions. First, they work with the back muscles to hold your body upright, supporting the weight of the chest cage, arms, and head. Second, they support and protect the spinal column. Third, they hold and support the internal organs. Fourth, they assist with the vital functions of breathing, waste removal, and reproduction.

The role of protecting organs is vital because, unlike the chest, which surrounds the lungs and heart with a bony cage, the abdomen is not protected by hard bone. Thus, the abdominal muscles must be strong in order to insulate and protect the organs—the stomach, liver, spleen, and intestines—that are located in the abdominal cavity.

Muscles of the Back

The muscles of your back also serve a number of functions. They help to control the movement of bending forward, by working as antagonists to the abdominal muscles. They allow the body to return to an erect posture after bending at the waist. They also work together with the abdominal muscles to maintain an upright posture. Back muscles fall into two groups. Superficial muscles of the back connect the ribs to the vertebrae, while the deep muscles of the back connect the vertebrae to one another.

The primary mover of the back and spine is the erector spinae. This is the muscle that causes you to straighten after you bend at the waist. This muscle group can be divided into three sections: the iliocostalis group, which assists with breathing by driving inhalation; the spinalis group; and the longissimus group. All three groups assist with extension or flexing of the vertebrae. The longissimus group muscles are concentrated in the lower back, while the other two groups are concentrated in the mid- to upper back.

The Muscles of Respiration

Three basic muscle groups drive the respiration process. The **diaphragm** is a sheet of muscle that defines the bottom of the chest cavity and separates it from the abdominal cavity. In addition to the diaphragm, there are eleven pairs of external intercostal muscles, located between the ribs and just below the skin, and eleven pairs of internal intercostal muscles, also located between the ribs, but below the layer of the external intercostals. You may be surprised to learn that the lungs themselves do not contain skeletal muscles. The role of the lungs in the inhalation and exhalation process is purely passive.

When the muscle fibers of the diaphragm contract, they cause the diaphragm to flatten and to lower slightly. This increases the size of the thoracic or chest cavity, which creates a vacuum because the chest cavity is a sealed cavity. The chest cavity is further expanded by the action of the external intercostals, which lift the ribs up and away from the lungs. The actions of the external intercostals and the diaphragm enlarge the chest cavity, creating a vacuum, and pulling on the tissue of the lungs. The lungs, which are flexible in a healthy human, stretch, opening the air sacs within and air rushes in from the mouth and nose to fill the space.

When the muscles of the diaphragm and external inter- costals relax, the weight of the chest cage causes the chest cavity to collapse inward, passively forcing out the air in the lungs. The role of the internal intercostal muscles is in forcible exhalation. For example, if you want to blow up a balloon, simply relaxing the diaphragm and the external intercostals will not provide enough force. The needed force comes from the internal intercostals, which contract, pulling the ribs toward each other and toward the lungs. This forces air out of the lungs and creates the air pressure you need to inflate that balloon or to blow out candles at a birthday party.

CONNECTIONS

In this chapter, we have explored the general anatomy of muscles. We have seen how muscles work in groups to create smooth fluid motions and how some muscles are designed for strength, while others are designed for delicate, accurate

SEE WHAT YOU HAVE LEARNED

In this chapter, we looked at selected muscle groups and how they function. Because of space and time, we did not look at the muscles of the arms and legs. Choose one of these sets of muscles and then find a good anatomy and physiology text (see the Bibliography for suggestions). Pick one of the arm or leg muscle groups and try to identify the agonists, antagonists, synergists, and fixators for each group. Ask yourself these basic questions to guide you:

- What is the primary function of the muscle group?

- How is this process reversed?

- Does the agonist need stabilizing and, if so, what muscles are likely to do this?

- Do the bones need stabilizing and, if so, by what?

You may be surprised at how much you already know.

movements. We have learned that while many muscles connect to bone, other can connect to softer tissues, allowing for subtle movements. We have learned that muscles usually work in opposing pairs and that these pairs not only can reverse the action of the opposite muscle group, but can also control the intensity of a movement. We have learned that some muscles are used to stabilize the movements of bone and other muscles, thus creating smooth, graceful movements.

Whether they are used for power, as one observes in a weightlifter, or grace, as one observes in a ballet dancer, the skeletal muscles allow us to move our body in an almost endless range of motions.

Glossary

Acromion The outermost and upward point of the shoulderblade where the collarbone attaches to the scapula.

Adenosine triphosphate The high-energy molecule that serves as energy currency for cells. Also known as ATP.

Agonist The muscle in a working group that exerts the majority of force.

Amino Acids The basic building blocks of proteins.

Amphiarthrosis Joint with limited freedom of movement. Plural is amphiarthroses.

Anemia A condition of the blood that results from the underproduction of red blood cells by the bone marrow stem cells.

Ankle bones A cluster of short bones at the junction of the lower leg and foot.

Antagonist The muscle in a working group that works in opposition to the agonist.

Antebrachium The forearm, extending from the elbow to the wrist.

Appendicular skeleton The bones of the shoulders, arms, and hands, and the hips, legs, and feet that are attached to the axial skeleton.

Arthrology The study of joints.

Articular cartilage Hyaline cartilage found where bones meet.

Articular facets Indentations in the radius that articulate with the bones of the wrist.

Articulate The point at which two bones come together.

Axial Skeleton The bones of the skull, thoracic cage, and spinal column.

Axis An imaginary straight line through the body.

Ball-and-socket joint A synovial joint consisting of a rounded end on one bone and a cup on the other. This joint allows a wide range of motion.

Base The end of the metacarpal closest to the wrist.

Belly The thickened middle area of many muscles.

Biaxial joint Joint that moves in two directions.

Bipennate muscles Muscles where muscle fibers connect on both sides of a central ligament.

Body The shaft of a metacarpal.

Bone marrow Soft tissue found within bones where blood cells are produced.

Bone marrow stem cells The cells in bone marrow that give rise to blood cells.

Bones The individual components that make up the endoskeleton.

Bony joint A fused joint between two previously independent bones.

Bony thorax The cage of bone, consisting of the thoracic vertebrae, the ribs, and the sternum (breastbone), that surrounds the chest.

Brachium The upper arm, extending from the shoulder to the elbow.

Brittle bone disease A disease that results in breaking of bones due to too little organic material.

Bursa Fluid-filled pads found between muscles or where tendons pass over bone. Singular is **bursae**.

Calcaneus The largest tarsal bone; also known as the heel bone.

Callus formation Development of hard tissue at a bone break or fracture that creates a temporary splint.

Canals Narrow tubes or channels in and between bones.

Capitate The most rectangular bone of the wrist.

Capitulum The condyle that contacts the radius.

Cardiac muscle Striated muscle found in the heart that is not under conscious control.

Carpal bones The bones of the wrist.

Carpus The wrist, consisting of eight small bones.

Glossary

Cartilage Softer skeletal tissue that does not contain high concentrations of calcium phosphate.

Cartilaginous joints Joints joined together by cartilage.

Cavity A hole or canal within the skull.

Centrum The porous core of a vertebra.

Cervical curvature The natural curvature of the cervical vertebrae toward the front of the body.

Cervical vertebrae The top seven vertebrae in the spinal column.

Chitin A structural molecule commonly found in the exoskeleton of insects and crustaceans.

Chondroblasts Cells that produce collagen.

Circular muscles See **Sphincter muscles**.

Coccyx Four fused vertebrae at the bottom of the spinal column that make up the "tailbone."

Collagen Fibrous protein that is the main connective tissue for bones and muscles.

Collagen fibers Tough connective tissue that contributes strength and resilience to bone.

Collarbone A thin, long bone that extends along the front of the shoulders.

Compact bone Bone containing lots of calcium phosphate. Also called dense bone.

Composite material A material made of two or more distinct components.

Conductivity The ability of muscle cells to communicate with neighboring cells.

Condyles Two small bumps on the distal (elbow) end of the humerus.

Condyloid joints A biaxial joint in the base of the fingers that allows the finger to move up and down and side to side.

Connective tissue Tissues made of mixtures of ground substance, fibers, and living cells.

Contractability The ability of muscle cells to shorten or contract.

Convergent muscles Fan-shaped muscles that are wide at their orgin and narrow at the point of insertion.

Coracoid process A fingerlike piece of bone on the scapula that serves as the attachment point for the biceps and other arm muscles.

Costal cartilage The pieces of cartilage that join the ribs to the sternum.

Cranial base The bottom part of the cranium, on which the brain rests.

Cranial cavity The chamber created by the cranial vault and cranial base.

Cranial vault A chamber surrounded by bone that stores and protects the brain.

Cranium The flat bones of the skull that surround and protect the brain.

Crural region The lower leg, between the knee and ankle.

Cuboid The largest bone in the distal row of the ankle.

Cuneiforms Three of the four bones of the distal row of the ankle (the fourth is the cuboid).

Diaphragm A sheet of muscle that separates the chest cavity from the abdominal cavity and assists in the breathing process.

Diaphysis The technical name for the shaft of a long bone.

Diarthrosis Joint with a high degree of freedom of movement. Plural is **diathroses**.

Differentiation Conversion of one cell type to another, as in the case of osteogenic cells giving rise to osteoblasts.

Digits The fingers and toes.

Diploe The middle, spongy layer of flat bones of the skull.

Elasticity The ability of muscle cells to stretch and return to the original shape and size.

Endomysium Fibrous connective tissue surrounding each individual muscle fiber.

Glossary

Endoskeleton A rigid, tough support structure on the inside of an organism.

Endosteum The membrane found inside the canals within bone.

Ends The two outside portions of a long bone (in its longest dimension).

Epicondyles Two bony complexes that flair out on the humerus just above the condyles.

Epimysium Connective tissue that surrounds entire muscles.

Epiphysis The technical name for the end of a long bone. Plural is **epiphyses.**

Erythrocytes Red blood cells.

Exoskeleton A rigid, tough, protective layer on the outside of an organism.

Extendibility The ability of muscle cells to elongate.

Extensors The muscles on the back of the neck that move the head up and backward.

False ribs Ribs that connect to the sternum indirectly through additional pieces of cartilage.

Fascicles Bundles of muscle fibers that work together to conduct functions

Femoral region The thigh of the leg.

Femur The bone of the thigh.

Fibrocartilage A type of connective tissue that is composed mainly of collagen fibers and is more fibrous than regular cartilage.

Fibrous actin Polymers of globular actin essential for muscle contraction.

Fibrous joint Joint where the collagen fiber from one bone extends and integrates into the adjacent bone.

Fibula The thinner bone of the lower leg that acts as a lateral strut.

Fissure Crevasse or gorge between bones.

Fixator The muscle in a working group that prevents a bone from moving in an unwanted direction.

Flat bones Bones that have a plate-like shape. They are narrow in one dimension and wide in the other two.

Flexors The muscles on the side of the neck that bend the head down and forward.

Floating ribs The two lower ribs, 11 and 12, that do not connect to the sternum.

Foramen magnum The hole in the base of the skull that allows access of the spinal cord to the brain.

Foramina Cavities that are classified as holes.

Forensic science The science of studying evidence from a crime scene in an effort to determine facts about the crime or the victim.

Fossa Ridge where bones come together, as in the cranial fossae. Singular is **fossae**.

Fusiform muscles Spindle-shaped muscles that are thicker in the middle than on the ends.

Fusion The joining together of two or more previously individual bones.

G actin See Globular actin.

Gelatinous bone marrow Bone marrow of unknown function that accumulates as we age.

Glenoid cavity The cup-like indentation where the ball of the humerus joins the scapula in the shoulder joint.

Gliding joints Synovial joints that overlap.

Globular actin Individual spherical-shaped molecules of actin.

Glucose A simple sugar that is the primary source of energy for the cells of the body.

Glycerol A three-carbon alcohol that serves as a source of energy for muscle cells.

Glossary

Glycoproteins Proteins modified with carbohydrates.

Gomphoses The joints that hold teeth into their sockets.

Granulation tissue Soft, fibrous tissue laid down during bone repair.

Ground Substance One of three components of connective tissue.

Hallux The large toe.

Hamate The wristbone at the base of the little finger.

Hamulus A hook of bone on the finger end of the hamate.

Hard callus A bony collar that forms during bone repair.

Haversian canal The opening in the middle of an osteon.

Head The end of the metacarpal closest to the finger or thumb.

Hematoma The site of internal bleeding.

Hematopoiesis The process by which bone marrow stem cells give rise to critical blood cells.

Hemopoietic tissue The tissue that gives rise to blood cells; also known as red bone marrow.

Hinge joints Synovial joints that move in one direction, usually with a 90–180° range of motion.

Hip bones A series of flat bones that are fused to form the pelvic girdle.

Humerus The long bone of the brachium, or upper arm.

Hyaline cartilage To come to come to come to come to come to come.

Hydroxyapatite The form of calcium phosphate found in bone.

Hyoid A U-shaped bone or series of bones located between the base of the tongue and the larynx.

Ilium The bone of the os coxae that is largest and closest to the head.

Insertion The more mobile end of a muscle.

Interosseous membrane The ligament that connects the radius to the ulna, and the tibia to the fibula.

Involuntary muscle Muscle that is not under conscious control.

Irregular bone Bones that do not fit into any of the other three categories.

Ischium The bone of the os coxae that is toward the back of the body and toward the feet.

Joint A point where two bones come together and are generally important in movement.

Joint capsule A fibrous capsule filled with synovial fluid.

Jugular notch An indentation in the bone at the top of the sternum.

Kneecap A sesamoid bone positioned over the knee joint.

Knuckle The rounded head of the metacarpal seen when a fist is made.

Lacunae Small spaces in bone that contain osteocytes.

Lap sutures Suture in which the bones have beveled edges that overlap.

Larynx The structure in the throat, made of cartilage, that houses the vocal cords. Also called the voice box.

Leach A process that results in the slow removal of minerals from bone.

Leukemia A cancer of the blood that results from the overproduction of white blood cells by the bone marrow stem cells.

Leukocyte White blood cell.

Ligament A piece of connective tissue that joins two bones and allows them to move relative to each other.

Lines of symmetry Imaginary lines that can be drawn through the body and that define centers for mirror images across the body.

Long bones Bones that are noticeably longer in one dimension than in the other two.

Lumbar vertebrae The section of vertebrae, usually 5 in number, between the thoracic vertebrae and the sacrum.

Lunate A moon-shaped bone of the wrist.

Glossary

Mandible The jawbone. The mandible serves as the anchor site for the lower row of teeth.

Manus The hand.

Mastication Chewing; the act of biting and grinding food in your mouth.

Maxilla The bone extending down from the nasal bone that serves as the anchor site for the upper row of teeth.

Medial epicondyle The epicondyle that houses the ulnar nerve and is commonly called the "funny bone."

Medial line The line of symmetry that runs from the center of the skull through the groin and that divides the body into left and right halves.

Medullary cavity Hollow cylinders in bone that contain marrow.

Meniscus A fluid-filled pad found between bones.

Menstruation A part of the normal reproductive cycle for women that results in the loss of blood as menses.

Metacarpals The five bones of the palm.

Mineralization The incorporation of calcium phosphate into connective tissue, resulting in the formation of bone.

Molt The periodic shedding of the exoskeleton in organisms to allow growth.

Monaxial joint Joint that moves only in one direction.

Motor unit All muscles fibers innervated or stimulated by a single neuron.

Multiaxial joint Joint that moves in many directions.

Multipennate muscles Muscles that are shaped like a group of several feathers joined by their quills at a single point.

Myoblasts Individual muscle stem cells that join together into muscle fibers.

Myocyte To come to come to come to come to come to come.

Myofibrils Protein bundles within sacroplasm.

Myofilaments The structural units of myofibrils.

Myoglobin A protein in muscle cells that stores oxygen for energy production.

Myology The study of muscles.

Nasal bone A ridge of bone at the top of the nose where the nose joins to the skull.

Navicular The triangular-shaped bone of the ankle.

Opposing pairs Muscle or muscle groups that reverse the movements of each other and that modulate the intensity of movements.

Orbits The two cavities within the skull that house the eyes.

Origin The relatively immobile or stationary end of a muscle.

Os coxae The fusion of three bones—the ilium, the ischium, and the pubis. Also known as the hip bone.

Ossa coxae Plural of **os coxae**.

Osseous tissue The technical name for bone.

Ossification The formation of hard bone.

Osteoblasts Bone-building cells.

Osteoclasts Bone-destroying cells.

Osteocyte Osteoblasts that have become trapped in the bone matrix and communicate between osteoblasts and osteoclasts.

Osteogenic cells Rapidly reproducing bone stem cells that give rise to osteoblasts.

Osteogenic layer The innermost layer of the periosteum that contains the cells for bone growth.

Osteon The outer portion of channels in bone, made of concentric rings. It has a cylindrical shape and is parallel to the longest dimension of the bone.

Pacemaker A special tissue associated with the heart that generates an electrical charge, causing the heart to beat.

Glossary

Parallel muscles Muscles with bundles of fascicles that are essentially equally wide at the origin, insertion, and belly.

Paraplegia A condition that results in paralysis of the lower limbs (usually from the waist down), most often resulting from a spinal cord injury.

Patella The technical name for the kneecap, a sesamoid bone.

Pectoral girdle The spinal column and upper portion of the thoracic cage that serve for the attachment of the shoulders and arms.

Pedal region The foot.

Pedicle A profusion of bone to which other bones, muscles, or tendons can articulate.

Pelvis (Pelvic girdle) The fused flat bones in the hips region that support the abdominal organs.

Pennate muscles Muscles that form an angle in relation to the tendon and pull at an angle, which means that the muscles cannot move their tendons as far as parallel muscles do.

Perimysium A connective tissue sheath surrounding fascicles.

Periodontal ligament A strong band of connective tissue that holds the tooth firmly to the jaw.

Periosteum The double-layered membrane that surrounds bones except for the joint surface.

Phalanges The bones of the fingers, thumb, and toes.

Phalanx The singular form of **phalanges** (the bones of the fingers, thumb, and toes).

Phalanx distal The finger or thumb bone closest to the tip.

Phalanx middle The bone in the middle joint of a finger.

Phalanx proximal The finger or thumb bone closest to the palm.

Pisiform A pea-shaped bone of the wrist.

Pivot joint A monaxial joint where a projection from one bone fits into a ring of ligament on the other.

Plane sutures Sutures occurring where two bones form straight, nonoverlapping connections.

Primary mover The muscle in a working group that exerts the majority of force.

Processes Projections of bone extending from vertebrae that are points of contact with other vertebrae and points of attachment for muscles.

Proteins Polymers of amino acids.

Proteoglycans Polymers of sugar modified with amino acids.

Pubis The bone of the os coxae that is toward the front of the body and toward the feet.

Quadriplegia A condition that results in paralysis of all four limbs, usually resulting from damage to the spinal cord in the cervical region.

Radius The smaller of the two long bones of the forearm.

Red bone marrow The tissue that gives rise to blood cells.

Remodeling The formation of repaired bone after the formation of hard callus.

Respiratory cartilage The cartilage of the pharynx, or voice box, that allows us to generate sound and ultimately speech.

Responsiveness The ability to muscle cells react to various stimuli.

Ribs A series of bones that curve around and protect the chest cavity.

Rickets A nutritional disorder causing bones to weaken and become deformed because of too little calcium phosphate.

Sacrum Five vertebrae just above the coccyx that fuse as we age and serve as the point of attachment for the hips.

Saddle joint A biaxial joint at the base of the thumb that allows the thumb to move up and down and toward and away from the fingers.

Sarcolemma The plasma membrane of a skeletal muscle cell.

Sarcomere A repeating unit of striated muscle fibrils.

Glossary

Sarcoplasm The cytoplasm of a skeletal muscle cell.

Sarcoplasmic reticulum The endoplasmic reticulum of skeletal muscle cells.

Satellite cell Muscle cell important in injury repair.

Scaphoid A boat-shaped bone of the wrist.

Scapula The triangular-shaped flat bone that, along with the collarbone (or clavicle) makes up the pectoral girdle. Also called the shoulder blade.

Serrate sutures Sutures connected by wavy lines, increasing the total surface of contact, and therefore making the suture strong.

Sesamoid bones Bones found within tendons that have a rounded end and a more pointed end.

Shaft The central portion of a long bone (relative to its longest dimension).

Sharpey's perforating fibers Little bits of collagen that anchor the periosteum to the surface of bone.

Shivering Involuntary spasms of muscles that generate heat.

Short bones Bones that are approximately the same length in all three dimensions.

Sinuses Small air-filled pockets within the skull that connect to the respiratory system.

Skeletal muscles The muscles that allow movement of the body.

Skull The series of bones that make up the head and jaw.

Smooth muscle Muscle associated with organ tissues that are not under conscious control.

Somatic motor neurons The nerve cells that signal muscle cells into action.

Sphincter muscle Muscle that forms a ring and allows constriction of an opening.

Spinal column The technical name for the backbone; actually a series of vertebrae assembled to form a column.

Spinal cord The nerve bundle that connects the brain to the rest of the body.

Spongy bone Bone tissue found in the ends of long bones and in the center of other bones that contains less calcium phosphate than compact bone.

Stem cells Cells that reproduce rapidly and give rise to multiple types of cells.

Sternum A flat bone in the front center of the chest that serves for attachment of ribs. Also called the breastbone.

Striated Having visible stripes or lines when viewed under the miscroscope.

Styloid process The projection of bone from the radius that is close to the thumb.

Sutures Fibrous joints that closely bind the adjacent bones and that do not allow the bones to move.

Symphases Cartilaginous joints joined by fibrocartilage.

Synarthrosis Joints designed not to move, or to move very little. Plural is **synarthroses**.

Synchondroses Cartilaginous joints joined by hyaline cartilage.

Syndesmoses Joints where two bones are joined only by a ligament.

Synergist Muscles in a working group that work additively with the agonist.

Synostoses Bony joints.

Synovial fluid A lubricating, cushioning liquid found in some joints.

Synovial joint A freely movable joint in which the bones are separated by synovial fluid.

Talus The bone of the ankle adjacent to the calcaneus.

Tarsal region The ankle of the leg.

Tendon A sheet or strip of tough collagen-containing connective tissue used to attach muscles to bone.

Glossary

Tendon sheaths Specialized bursae that wrap around tendons.

Terminal cisternae Small sacs in the sarcoplasmic reticulum that are storage sites for calcium ions.

Thoracic vertebrae The section of vertebrae, usually 12 in number, below the cervical vertebrae, to which the ribs attach.

Tibia The larger weight-bearing bone of the lower leg.

Trapezium A near-circular bone of the wrist.

Trapezoid A four-sided geometric shaped bone of the wrist.

Triquetrum A triangle-shaped bone of the wrist.

Trochlea The condyle that contacts the ulna.

Trochlear notch The point of articulation between the humerus and the ulna.

Tropomyosin The protein that makes thin filaments in muscles.

Troponin A calcium-binding protein in bone.

True ribs Ribs that connect directly to the sternum through short pieces of cartilage.

Trunk The central core of the body, consisting of the chest and abdomen.

Tympanic membrane The technical name for the eardrum; a membrane that blocks the entrance to the inner ear and is important in hearing.

Ulna The larger of the two long bones of the forearm.

Ulnar nerve The nerve that crosses the medial epicondyle and is the site for stimulation of the "funny bone."

Ulnar notch The groove in the radius where the ulna contacts the radius.

Unipennate muscles Feather-shaped muscles where all muscle fibers attach to one side of a tendon.

Vertebra An irregular bone that helps form the spinal column, or backbone. Plural is **vertebrae**.

Vertebral column The arrangement of the vertebrae into a structure (also called the spinal column) of the axial skeleton.

Vertebral foramen The hollow circle in each vertebra that when assembled with the other vertebrae make up the tube for the spinal cord.

Vestigial structure A remnant structure in the body that no longer serves its original purpose and has limited utility.

Voice box The larynx, a structure involved in sound production and speech.

Volkmann's (perforating) canals Small canals at right angles to the Haversian canals of the osteon.

Voluntary Under conscious control.

Voluntary movement Movement of a muscle or limb that is under conscious control.

Wrist bones A cluster of short bones at the junction of the lower arm and hand.

Xiphoid process A separate, diamond-shaped flat bone at the lower end of the sternum.

Yellow bone marrow Bone marrow rich in fatty tissue, but containing few if any stem cells.

Zone of bone deposition The site in growing bone where lacunae begin to break down and chondrocytes begin to die off.

Zone of calcification The site in growing bone where mineralization occurs.

Zone of cell hypertrophy The growth zone of bone where chondrocytes stop dividing.

Zone of cell proliferation The site toward the center of a bone nearest the zone of reserve cartilage.

Zone of reserve cartilage The region farthest away from the center of the developing bone.

Bibliography

Alcamo, E.I., and I.E. Alcamo. *Anatomy and Physiology the Easy Way.* Hauppauge, NY: Barron's Educational Series, Inc., 1996.

Anatomy and Physiology for Dummies. Hoboken, NJ: John Wiley and Sons Publishing, 2002.

Anatomy and Physiology Made Incredibly Easy. Philadelphia: Lippincott Williams & Wilkins, 2000.

Marieb, E.N., J. Mallatt, and M. Hutchinson. *Human Anatomy and Physiology*, 6th ed. San Francisco: Benjamin-Cummings Publishing, 2003.

Muscalino, J. *The Muscular System Manual: The Skeletal Muscles of the Human Body.* St. Louis: Elsevier Science Press, 2003.

Scanlon, V., and T. Sanders. *Essentials of Anatomy and Physiology*, 4th ed. Philadelphia: F.A. Davis Publishing, 2002.

Stone, R.J., and J.A. Stone. *Atlas of Skeletal Muscles.* New York: McGraw-Hill Company, 2002.

Further Reading

Alcamo, E.I., and I.E. Alcamo. *Anatomy and Physiology the Easy Way.* Hauppauge, NY: Barron's Educational Series, Inc., 1996.

Anatomy and Physiology for Dummies. Hoboken, NJ: John Wiley and Sons Publishing, 2002.

Anatomy and Physiology Made Incredibly Easy. Philadelphia: Lippincott Williams & Wilkins, 2000.

Aurou, E., and A.M. Tenllado. *Skeletal System.* Broomall, PA: Chelsea House Publishers, 1995.

Avila, V. *How Our Muscles Work.* Broomall, PA: Chelsea House Publishers, 1994.

Barnes, K., and S. Weston. *How It Works: The Human Body.* Barnes and Noble Publishing, 2000.

Parker, S., and C. Ballard. *Skeleton and Muscular System.* Redwood City, CA: Raintree Publishers, 1997.

Silverstein, A., R. Silverstein, and V.B. Silverstein. *The Muscular System.* Brookfield, CT: Millbrook Press, 1995.

Simon, S. *Bones: Our Skeletal System.* New York: HarperCollins Publishing, 2000.

White, K. *Muscular System (Insider's Guide to the Body Series).* New York: Rosen Publishing Group, Inc., 2001.

Websites

American College of Sports Medicine,
 Information on Muscle and Bone Injury
 http://www.acsm.org/index.asp

American Orthopaedic Association
 http://www.aoassn.org/

Medline Plus, a Service of the National Library of Medicine
 and the National Institutes of Health
 http://www.nlm.nih.gov/medlineplus/medlineplus.html

National Institute of Arthritis and Musculoskeletal and
 Skin Diseases (NIAMS), National Institutes of Health
 http://www.niams.nih.gov/

National Institutes of Health
 http://health.nih.gov/

National Osteoporosis Foundation
 http://www.nof.org/

Unit (metric)		Metric to English		English to Metric	
LENGTH					
Kilometer	km	1 km	0.62 mile (mi)	1 mile (mi)	1.609 km
Meter	m	1 m	3.28 feet (ft)	1 foot (ft)	0.305 m
Centimeter	cm	1 cm	0.394 inches (in)	1 inch (in)	2.54 cm
Millimeter	mm	1 mm	0.039 inches (in)	1 inch (in)	25.4 mm
Micrometer	μm				
WEIGHT (MASS)					
Kilogram	kg	1 kg	2.2 pounds (lbs)	1 pound (lbs)	0.454 kg
Gram	g	1 g	0.035 ounces (oz)	1 ounce (oz)	28.35 g
Milligram	mg				
Microgram	μg				
VOLUME					
Liter	L	1 L	1.06 quarts	1 gallon (gal)	3.785 L
				1 quart (qt)	0.94 L
				1 pint (pt)	0.47 L
Milliliter	mL or cc	1 mL	0.034 fluid ounce (fl oz)	1 fluid ounce (fl oz)	29.57 mL
Microliter	μL				
TEMPERATURE					
$°C = 5/9 (°F - 32)$		$°F = 9/5 (°C + 32)$			

Index

Index

Index

Index

Picture Credits

page:

About the Author

Dr. Gregory J. Stewart completed his Ph.D. in microbiology from the University of California at Davis. His post-doctoral training was conducted at Exxon Research and Engineering and at E. I. DuPont de Nemours and Company. He spent seven years at the University of South Florida as an Assistant and Associate Professor of Biology. In 1993, he joined the Biology Department of the State University of West Georgia (then West Georgia College), where he served as department chair for eight years. After one year as Assistant Dean of Arts and Sciences at West Georgia, Dr. Stewart accepted a senior fellowship with the Bureau of Arms Control at the U.S. Department of State, where he is currently employed. His duties at the State Department include coordination of the Biological and Toxins Weapons Convention (a treaty) and serving as an advisor to all bureaus and other federal agencies on issues related to microbiology, biological weapons reduction, bioterrorism/counter-terrorism threat reduction and response, and nonproliferation of weapons of mass destruction. He also serves on the National Institute of Allergy and Infectious Diseases' biodefense vaccine development review panel and on the Small Business and Innovative Research panel for biodefense response, as well as serving as ad hoc reviewer for several other funding agencies and professional journals. He is the author of more than 30 scientific publications. Dr. Stewart lives with his wife, Patricia, and is the stepfather of three children and the grandfather of one.